Crossroads of Culture

Crossroads of Culture

A Study in the Culture of Transience

Pulin K. Garg
and
Indira J. Parikh

Sage Publications
New Delhi/Thousand Oaks/London

First published in 1995 by

Sage Publications India Pvt Ltd
M-32, Greater Kailash Market-I
New Delhi 110 048

Sage Publications Inc
2455 Teller Road
Thousand Oaks, California 91320

Sage Publications Ltd
6 Bonhill Street
London EC2A 4PU

Published by Tejeshwar Singh for Sage Publications India Pvt Ltd, typeset by Line Arts, Pondicherry, and printed at Chaman Enterprises, Delhi.

Library of Congress Cataloging-in-Publication Data

Garg, Pulin K., 1927–
Crossroads of Culture: a study in the culture of transience/
Pulin K. Garg and Indira J. Parikh.
p. cm.
Includes bibliographical references.
1. Social change—India. 2. Rural development—India.
3. Technology—Social aspects—India. 4. Culture. I. Parikh.
Indira J. II. Title.
HN683.5.G383 1995 306′.0954—dc20 95–7274

ISBN: 0–8039–9234–3 (US-hb) 81–7036–463–9 (India-hb)
0–8039–9235–1 (US-pb) 81–7036–464–7 (India-pb)

Sage Production Editor: *Niti Anand*

Dedicated to

Gaurav K. Garg
and
Sushrut J. Parikh

Representatives of children of two cultures
and creators of culture of tomorrow

Contents

Acknowledgements

The book owes its existence to hundreds of young men and women in the age group of twenty to twenty-five, from all over India, who joined us in exploring the dynamics of their identity. The setting of these explorations were two courses conducted year after year since 1972 at the Indian Institute of Management, Ahmedabad. These courses constituted a learning situation where the participants shared their experiences of growth in the settings of family, education institutions, neighbourhood, society at large and culture in general. We acknowledge the valuable gift of these young men and women. Their participation helped us outline the broad contours of the interaction between society and culture and eventually capture the milieu of their growth.

Later this mode of creating a learning setting was extended as the basic core of work at the Indian Society For Individual and Social Development (ISISD). For the last thirteen years ISISD has organised programmes during summer as well as other times where a community of persons has come together to explore its identity processes. These groups have included children from eight to thirteen years of age; adolescents from fourteen to nineteen; young adults from twenty to twenty-five and adults comprising management professionals, teachers, social workers and women who run their own households. In their explorations they too shared their experiences in different settings. They shared their joys and anguish; their wellbeing and pathos; their conflicts of role and self and their transactions with systems and structures. They voiced their hopes, dreams and aspirations and their fears, frustrations, resentments and sense of helplessness. They talked of their efforts to keep themselves mobilised to be effective and efficient individuals. Their participation was open and candid. In the reliving of their experiences they traversed a large canvas of Indian society.

We consistently joined them in exploring our own processes of identity but also undertook to evolve frames which would integrate such a wide variety of experiences. From these frames we began evolving constructs to conceptualise the experiences. A whole host of colleagues within ISISD often joined us in reviewing their own experiences of similar work. We periodically reviewed our emergent frames, constructs and conceptualisation with them in order to formulate a systematic perspective to support our continuing work. We acknowledge the contribution of more than two thousand participants through these thirteen years. The colleagues at ISISD are many. Sushanta Banerjee, Shyamal Gupta, Gopal Khandelwal, Suprio Chowdhary, Zahid Gangjee, P.M. Kumar and Subrata Roy are some of those who have been closely associated with the process. Besides them we wish to acknowledge the contribution of Professors J.V. Singh (Wharton School), V.K. Narayanan (Kansas University), Vipin Suri (Management Consultant) and Kapil Malhotra of Vision Books. Our special thanks go to Smita Suri and Seetha Ananthasivan who went through the manuscript twice and gave their comments.

The book has been long in writing. At no time were we satisfied that we had captured the dynamics of identity formation sufficiently. One version was ready in 1976 and another in 1978. It was rewritten in 1982 and then again in 1986. It was finally recast in 1990. During these fourteen long years a host of our friends pushed us to give it a final shape. Many of them have extended support by giving us their time and affording us facilities. Mr. A.K. Gupta, General Manager, Hotel Clark's Amer offered us a week-long stay at his hotel so that we could shut out the many competing demands on our time. Once again Vipin Suri, Manager, Hotel Searock (Welcome Group), and his wife, Smita Suri, provided us the hospitality of their apartment as well as a working suite in the hotel. Their contribution indeed is far beyond the actual gift of space. They kept our morale high.

Our unqualified thanks go to a series of secretaries at IIMA who through the years typed the various versions. Among them are Mr. V. Jagannathan, Mr. R. Venkatesan, Mr. Subramanian and Ms. R. Usha. The final version was put together through the word processor by Ms. Falguni Patel, the administrative secretary and an associate of one of the authors. Mr. Tamil Selvan, Mrs. Nayana Shah, Ms. Nina B. Muncherji, and Ms. Babita Mathur, research staff in the Organisational Behaviour area at the Indian Institute of Management, Ahmedabad, provided referencing support. Mrs. Nayana Shah read and reread the manuscript for editorial purposes. Our thanks to all of them.

During the periods when we found a sense of futility overwhelming us, when we felt depressed and angry with ourselves for not being able to get a grip on things, when we were moping around, Nalini Garg and Jitendra C. Parikh saw us through our agony and encouraged us. They put up with our moods and turbulence and persistently motivated us. From time to time, we used them as sounding boards for sorting out our confusion and helplessness. They made us realise that in undertaking this project we ourselves were going through the various cultural processes of non-action and postponement of self—the processes which have become deeply ingrained in the Indian psyche. Their contribution and support is of immense value. We acknowledge that without their active encouragement of our efforts, we may not have been able to draw the boundaries and complete the writing.

Tejeshwar Singh, who has also published two of our previous books, spent a great deal of time pushing us to complete the book. We appreciate his being patient with us. We also deeply value our long association with him.

Pulin K. Garg
Indira J. Parikh

One

Preamble

This book attempts to explore the nature, quality and characteristics of the changes occurring in the agrarian ethos of India and the transience that is the experiential reality of industrialisation in India. Transience is both a state and a process (Parikh, Garg and Garg 1988). Transience emerges in a society when the individuals in it are exposed to two distinct types of ethos about the nature of collectivity and man's relationship to his collectivity. In Indian society, the collectivity had a traditional, agrarian ethos which provided continuity. The subsequent encounter with the Western ethos generated discontinuity in living processes. This creates the culture of transience. There is enough evidence to suggest that the culture of transience prevails in all post-colonial societies (*ibid*).

Let us look at the concept of culture. At the content level culture encompasses everything from do's and don'ts, beliefs, values, myths, folk tales, rituals, institutions, customs and religion. It spans all aspects of living. As such, cultures sustain and bind together divergent elements. At the process level culture holds the throbbing, seething dynamic and the stagnant, decaying passive contents to create a life space. Culture as a process then tends to hold diverse elements in the life space of man in correspondence, congruence, convergence and coherence to create a rhythm and theme of life which individuals enact (Parikh, Garg and Garg 1988, 50).

The individual and the collectivity then enact the contents and processes of the culture in divergent ways. Our encounter and understanding of the culture of transience in India has emerged from various sources. We have been involved in developmental action in organisations, in the process of culture and institution building, in helping individuals design effective roles in their life space, and with a whole set of young adults in systems of higher and professional education. More specifically, over the last fourteen years these interactions have taken place in group settings to explore the issues of role taking and the meaning of being members of the family system, the organisation, and society at large. We have explored their world-views, the present context they experience, and the future they visualise for themselves, their families and society. During this exploration they have articulated feelings, perceptions, and problems of personal choices which are not normally shared in public. Inevitably, these explorations have reflected patterns of authority relating to social norms, to values perceived while growing up, and the processes which shaped them as individual role holders.

The data generated from these explorations in group settings and interviews suggested two unique patterns. One pattern reflected the processes of being and becoming and the crystallisation of an Indian identity (Garg and Parikh 1976, 1980). In our attempt to portray these processes, we found that the Indian youth today is a child of two cultures: Indian and Western. Young people internalise contradictory and conflicting world-views, lifestyles, and modes of relationships with others; caught between an agrarian, traditional ethos and the industrial, Western ethos the Indian identity becomes fragmented.

This fragmentation is a part of the failure of Indian society and culture to synthesise the two dominant types of ethos. The youth has become a victim of 'double bind'. As Gregory Bateson uses it, double bind implies a process where contradictory messages converge on the individual from either two distinct sources or from the same source. We have hypothesised that the Indian child internalises significant models of contradictory thought and action from the processes of the various systems of which he is a member.

The second pattern reflected the elements of culture and society being experienced, given meanings to and the resultant action choices. This book is our attempt to put together and analyse the systemic data from these explorations in order to delineate the Indian identity and the emotive and cognitive maps of Indian society as viewed by the participants. Indian culture can be seen as a living process assimilating various strands of thought and lifestyle which it encounters. The process has created a rich collage of India exemplifying 'unity in diversity and diversity in unity'. However, the historical encounter with Western technology and industrialisation, together with its ethos, has been drastically different in nature and intensity than any earlier encounters. When Indian society was confronted with the Western ethos, assimilation and harmony were not the outcome of the encounter. The individual and the system became fragmented into a logical, rational and cognitive domain of thought and action on the one hand and an emotional, psychological, and relational domain on the other. The individual's thoughts and feelings about the situation could not converge so as to help him make a clear choice of action.

This fragmentation in the Indian identity manifested itself in the diverse areas of home, education, neighbourhood and formal work organisations. This fragmentation was expressed and responded to in many ways.

Fragmentation of Emotive and Cognitive Life Space

Until the 1940s Indian society managed this encounter with the Western ethos and the resultant fragmentation by dividing the life space into two zones—the private and familial, and the public and communal. There was some continuity in the familial and the relational modes but new public and community institutions were created. These institutions revolved around peer group activities such as scouting, sports meets, debating, dramatics and other interests. Organisationally, rules, regulations and formal structures were designed in order to achieve objectives and goals which were all based on the Western ethos. However, in actual, everyday living the substratum of familial and affiliative processes and structures of an agrarian society operated. This meant that the agrarian affective and affiliative ethos was extended and carried over into the neighbourhood. Schools and many other secondary systems began to be identified and included as 'my' community. All institutions were built in the public and community zone by the reinduction of the processes of the Indian ethos. The written constitutions with rules and regulations designed on Western models retained their legal, formal tonality while in expression and action the emergent actions reflected the agrarian ethos. On the other hand, Western notions of democracy, justice and fair play were introduced into institutions of higher learning.

Up to the 1940s, this division between the public and the private worlds helped earlier generations and their institutions to achieve a semblance of stability and purpose. They operated with Western logical and rational forms in work and knowledge systems but remained rooted in the emotional and relational ethos of India. They managed this dichotomy successfully. The generations reaching adolescence after Independence were exposed to a scientific and technological milieu. They were impatient to make up for past deprivations, and to catch up with the technological and industrial ethos of the West. They had new aspirations and ambitions. They found it difficult to accept the inconsistencies, hypocrisies and contradictions of the older generation and social institutions. Later, this feeling of alienation developed into scepticism and/or cynicism.

We carried out extensive dialogues to discover how the new generation experienced the socio-cultural systems and their roles in it. These dialogues can be grouped into several themes.

Lack of Flexibility in Exercising Choices

Many individuals said that their roles in the external and internal family systems did not correspond. Within the educational system they had moved from one academic achievement to another. Their choices could be seen as logical, appropriate and relevant. However, looked at objectively, their choices had followed the beaten path and were largely determined by social, familial and institutional influences. They had rarely, if ever, made a personal choice. (They had not been able to withstand either social processes or peer pressure.) They had allowed themselves to drift and move with the herd.

The young adults appeared to have confidence in their abilities and future prospects, but were plagued with doubts and a sense of inadequacy. Although their positive achievements gave them social and academic status, they carried a poor sense of self-worth. Some also reported that they did not feel they had the option of making a personally relevant choice and acting autonomously, as they lacked the guts to do so. They ended up conforming. For many of the individuals the dominant feeling in retrospect was 'I always knew I would succeed', although during their actual transaction with the world they were unsure.

The young adults experienced much turmoil at the threshold of their careers. They applied for innumerable jobs. They sat through endless interviews. They were at the crossroads. One path led to multinationals, to large, professionally managed organisations, which offered hoof money, opportunities for going abroad, membership to an exclusive group and high resale value. These were the factors that counted in the market. But the choice involved being just one among many, beginning at the lowest rung of the ladder, doing work at a much lower level than one had been trained for, and living a lonely life in a metropolis. The other road led to a medium organisation, possibly family owned. The beckoning factors were less competition, and the opportunity for rising in the hierarchy quickly, using knowledge to bring about change, and finally, the chance to work in close relationships. However, this path carried fears of a possibly authoritarian set-up, an undefined level of professionalism, a lower salary, and very little opportunity for going abroad. In both choices, there were anxieties about being able to settle down and the necessity of moving from job to job.

For the managers the turmoil had different dimensions. They discovered that in order to succeed and achieve the desired social status symbols, they had postponed doing many things which would have made their lives more meaningful. They recognised that they had become estranged from their spouses and children. They brought only their 'products' to their families but withheld themselves. They experienced great stress. They also sensed that there was little challenge left, and their choices were resolved to a limited few. They could move laterally with a slightly better salary, go abroad for more money, build a power base, seek temporary emotional linkages outside marriage or endow themselves and their family with symbols of higher social status. Basically, they were incapable of disengaging themselves from the pattern and lifestyle they had got into. Only a few had the courage to start out on their own.

To us, the choice facing these individuals, the young adults and the mature managers is analogous to the prince in the folktale who, burdened and oppressed by the heavy burden of his traditionally bound role in society, walked out one day on a journey of self-realisation and becoming.[1]

[1] There are various versions of this folktale. The most popular one in the North India is that of a prince who is the youngest child of the family. He leaves his kingdom in order to discover his own strength. He has many adventures, crosses a forest, and reaches a crossroad where a witch sits. He seeks her advice. She tells him, 'The road to the right is long and easy, but the road to the left is short and dangerous. A demon controls the short road. If you go on this road, it would be best for you to address the demon as *mama* (maternal uncle). You have to eliminate him by discovering the secret of his life before you can go across.'

Like the hero in the folktale the individuals have crossed the forest and stand at crossroads. The choice is theirs, so is the dilemma. It is easier to follow the beaten path, the road trodden by many; easier to be in the good books of people who matter in the system. To take a new road and create a new space and identity for oneself is both difficult and frightening.

Many participants dreamed of creating a world of their own. However, such a choice was fraught with fear. It lacked security. It demanded faith in one's own abilities and potentiality. It implied a trust in one's ability to put in committed effort, and work to overcome possible constraints. It demanded strength to be self-reliant, and readiness to pay the price for the choice made. It generated anticipatory anxieties about failure, loss of face and status, which those who chose otherwise would not face. Essentially this choice implied walking alone and braving the critical scrutiny of everyone around. Some of the young adults eventually did make this choice but it took them from one to seven years. Some managers also left their jobs and ventured into entrepreneurships.

For the young adults, the alternative of making socially desirable choices ensured acceptance by others. It provided primary props to sustain their induction into the adult world. It implied taking on a performer's role and leaving the management of the backstage and the wings to others. Responsibility for failures in performance was then displaced to the organisation and its members. It was a safe path.

Sets of Choices

Choice making invariably generates conflicts and dilemmas. The dilemmas arise out of conflicting priorities regarding material rewards

The prince makes the choice and takes a short but dangerous road. He comes across the castle of the demon where a princess is held captive. The prince lives under the scrutiny of the demon and discovers that the demon's life is bound by magic with the life of a parrot which is kept in a secret chamber.

The prince devises ways first to discover the secret chamber and then to kill the parrot with the help of the captive princess. Thus, in one act he becomes a hero, killing the demon and receiving the love of the princess. He returns as the total man to his kingdom.

It is an interesting story. The parrot symbolises the comforting ethos which makes it difficult for the young prince to be an adult in his own right. His killing of the parrot is a symbolic act of destroying the overwhelming and oppressive 'shoulds' which allow him no choice but to be a carbon copy. The princess is his own ego, with which he unites to find his freedom and be a adult in his own right.

and expectations and the contradictions between the traditional and modern. The traditional focus of identity drew them towards the emotive map of Indian culture which emphasised familial affiliative processes, interpersonal interdependence and a certain degree of affinity with the system. The modern foci of identity impelled them to seek an individualistic, achievement-centred stance which would enable them to carve out a singular role and space in life for themselves.

These foci implied different lifestyles—either modern or traditional or an uneasy blend of the two. The response to these dilemmas manifested itself in many spheres of life.

Lifestyles

Modern lifestyles meant earning a high pay, working for a multinational firm or big business house, and Western ways of social living. This implied drinking, going to clubs and parties, and being visible in the settings of conspicuous consumption. Such a style displayed the need to project an image of a well-to-do, successful individual, who could afford the graces and conveniences of a sophisticated, Western culture. It implied an elaborate Western style arrangement of space in the home with Western furniture and decor. It meant the inclusion of Western dishes in food, and observing festivals as social occasions while ignoring their religious and ritualistic aspects. Most individuals tended to choose their own spouses. They desired their wives to be socially active or to work, and wanted them to be well-groomed and modern in their outlook. Children were sent to private schools or other elite institutions.

Traditional lifestyles meant the average, middle class, urban or semi-urban mode of living. Men and children could wear Western clothes, but the wife continued to wear Indian clothes. It often implied using functional Western furniture but with Indian decor. Importance was given to social interaction with neighbours and one's own community, and a ritualistic celebration of festivals. Eating out included the family, and popular 'desi' places having a regional flavour were patronised.

The choice of lifestyles was not determined by monetary considerations, but resulted from psychological preferences for certain lifestyles. Even when the question of marriage arose and these young adults wanted to choose their spouse themselves, the parents were required to screen the potential partners. Love could be dreamt of but not necessarily actualised in marriage.

Only a few were clear and sure while choosing between the two lifestyles. Most of them wanted a mix of both. Among those who wanted a totally modern lifestyle, the basic conflict of location of identity could be observed. A good number of them wanted only the trappings of a modern lifestyle, while retaining beliefs and family relationships moulded by the traditional Indian ethos.

Location of Identity

What did it mean to have a modern identity? It meant primarily to be rational, to have a scientific approach, and hold empirally tested beliefs, with little room for faith. Faith was considered to be synonymous with superstition. Behaviour based on feelings was sentimental and therefore to be avoided. Education was purely techno-informative and a means of acquiring control over large segments of worldly resources. Completing tasks and obtaining recognition for doing so were not enough to count as achievement. Moving upward in status, acquiring power, and even greater control of worldly resources were also essential. To be modern also implied confining one's cultural life within the chosen system.

The data from these explorations suggested that the young adults were action-oriented which provided social visibility and monetary rewards. They lived a very narrow intellectual life. Most of them knew nothing of serious literature, either in their own mother tongue or in English. They were also not well acquainted with classics of modern knowledge, sociology, philosophy or history. Their knowledge was confined to science and technology. What they knew of humanities was through tertiary sources or from encapsuled editorial and journalistic opinion-making articles in magazines. Their reading, in their growing years, was confined to authors like Enid Blyton and various kinds of comics; and as adolescents to current bestsellers of crime and espionage and such like.

Essentially, being modern was to make a distinct break with the ethos prevalent prior to the 1940s. These individuals represented the first generation which had had a techno-informative education. National boundaries did not confine them. They saw the whole world as their playground, were willing to go where their talent would be best used and where they would experience the fewest constraints in their chosen lifestyle. This implied a certain degree of disengagement in their primary relationships, among other things.

Both men and women, however, retained a very traditional orientation towards the gender role differentiation in marriage. The man displayed superiority in status and dominance in decision-making, thought, feeling and action. He saw himself as the pivot of socio-economic existence. Women in his universe were subservient and dependent. In our samples women intellectually protested against this orientation but surprisingly gave in to it on an emotional plane.

Cognitively, they resented inequality and argued for equality. They wanted to be educated, behave as modern wives and mothers but simultaneously wished to retain their traditional dependence on men, wanting them to dominate. Their modernity and education were to be used only for contributing to their children's development, and the maintenance of their new social status.

Middle-class society, specially in the United States, was a reference group for all economic, technical, and scientific progress and development. The emphasis of these young Indians was on borrowing and adopting the Western lifestyle. It meant operating within a comparative framework of accepted inequalities. The Indian reality could not be perceived on its own terms. Anything Indian had to be reaffirmed and validated by the Western intellectuals and elite, before it could be accepted or adopted. The personal experiences of the Indians, for instance being cured of illness by ayurvedic medicine, was treated as either a fluke or an exception. Even contrary personal experience did not shake their belief in the West as a paradigm of superiority. This was true of Indian classical music, Indian dance and art forms. The exception was Indian cinema, which anyway borrowed heavily from the West.

On the other hand, having a traditional identity meant retaining faith as one of the basic elements which gives meaning to life, reading literature and scientific works written in Indian languages, while not excluding Western scientific works from one's reading list. Attempts were made to establish a correspondence between the newly discovered principles of Western knowledge with an Indian interpretation of the phenomena. Other characteristics of the traditional identity included the idea that being sentimental in relationships was not taboo. To a certain extent sentiment based on normative behaviour was accepted, but too obvious a display of it made them uncomfortable. It supported a tacit acceptance of the mother crying while saying goodbye to her child at the end of a holiday or calling upon one's parents' acquaintances in the city where one was being educated, without feeling compelled to do so. Education and achievement were valued as much for personal success as for the credit it brought to the family. Looking after the wellbeing of the family was still considered a virtue.

Being traditional in identity also meant condemning some social customs and practices and cultural modes of Western society. The norms for these individuals were taken from Indian society and culture. They were willing to take a reformist view and condemn the custom of dowry, the caste system and other social evils. However, they were not willing to accept the social and cultural modes of the West as indications of progress and development. Western modalities of economic, technical and scientific aspects of life were acceptable, but their relevance and adoption were questioned. Developing regional languages and imparting instruction in these was advocated but not practised. A good knowledge of English was seen as a necessity. In a sense, there was a sectoral differentiation in the point of reference, i.e., for the economic and technical development the West was the model, and for social and cultural processes the Indian tradition was the best.

The men and women with whom we talked fell into the following two patterns:

1. Modern in identity and traditional in lifestyle.
2. Modern in lifestyle and traditional in identity.

Very few belonged to a possible third category of being traditional in lifestyle as well as in identity. Over the years the nature of social change, its fast pace and direction suggests an increasing number are being attracted by a modern lifestyle and a modern identity.

The fragmentation resulting from these asymmetrical combinations of modern identity and traditional lifestyle on the one hand, and traditional identity and modern lifestyle on the other, could largely be attributed to the family culture. The children from families which had acquired a high socio-economic status after the Second World War often grouped themselves in the modern identity and the traditional lifestyle category. However, they most actively indulged in conspicuous consumption.

Those who more often categorised themselves as traditional in identity and modern in lifestyle either came from families with a long professional tradition enjoying a high cultural status in society, or from those which had been active in the struggle for freedom and cultural renaissance before Independence.

The above classification is reflective of today's dilemmas in Indian society, resulting in young people having aspirations which often lead to conflicts in relationships. They feel the pressure of parental expectations and feelings and other constraints which make choices difficult.

At another level the individuals are involved in two major dilemmas reflecting authority and interpersonal relations.

Nature and Location of Authority

Relating with authority was the core issue around which intensities of emotions emerged. Relating to fathers and mothers brought forth a range of emotive responses, from concern to contempt, from involvement to indifference, from anxiety to pity, from love and respect to resentment and anger. There was a struggle to relate to them as people and not just as parents. Relation to authority which was personal, emotional and encompassing left the individuals with dominant, undifferentiated modes of response and generated turmoil. In the academic institutions response to authority was largely indifferent, hostile or negative. Only a few teachers commanded respect by being accepted as role models. The rest received a contemptuous tolerance and/or indifference. Organisationally, the issue of authority revolved around the questions: Should authority be shared by the people in the system or should it be held by one person? Should the individual have some self-authority (autonomy) or should his authority always be sanctioned from outside?

This issue has plagued the Indian identity. A logical, rational and shared concept of authority has not found enduring roots. The culture of transition has created intensities and devalued the experiences of the older generation. The education system values information and knowledge as derived from the media and Western sources. It also tends to follow what is current in belief or behaviour patterns.

Nature and Quality of Interpersonal Relations

The issue of relationships both with the family as well as peers generates similar emotional intensities. The culture of transition means that all the traditional modes of relating become obsolete. The wisdom and experience of the older generation is replaced by inadequacy in making choices in the education systems. The younger generation is well versed in scientific terminology, general information and the environment. Younger women are better educated than their elders, and have acquired the know-how of managing people and the world around. Younger individuals are more familiar with technology. All this has led to diffidence, tentativeness and uncertainties in relationships. Working out a close and intimate relationship is fraught with possessiveness, control and ego conflicts. The problems of relating to, and between, given relationship and voluntary relationships create stress. The issue can be articulated as: Should the demands of relationships in which one is born in (involuntary relationships) outweigh the demands of self-created and self-chosen relationships (voluntary relationships)? If so, how should the individual draw a boundary between the two?

This problem was perhaps not felt so poignantly as the struggles for creating relationships at the peer level. The process of forming peer relationships and attempts to create intimacy and closeness were the contexts in which most of the struggles for identity locations and choice of lifestyles became acute. It was during these struggles that the individuals tried to look at their sense of belonging, their roots, and the meaning they wanted to give to their lives.

Definition and Perspective of Progress and Development

The current generation is exposed to diverse modes of living from across the world. In their exploration of definitions and perspectives they visualised for the country's growth and development, there were mixed responses. Most individuals were acutely aware of a large number of social, cultural, economic and political dysfunctionalities in Indian society. The Indian citizen of today is systematically overexposed to all the negative aspects of his own society and culture. The result is that the few positive aspects he is exposed to appear irrelevant to him. He cannot objectively appraise the achievements of the country and culture.

In contrast, most individuals were acutely aware of the great economic and technological strides reflected in the standard of living of an average citizen of some Western nations. The younger generation wanted Indian society to follow suit. Most of them ignored the fact that adopting the technological and economic modes would bring Western social and cultural processes as well into the family and social living. They believed that the structure of the techno-economic system can be divorced from the socio-emotional living process, and can be bounded within the field of economic activity. They failed to appreciate that such a selective adaptation and assimilation is impossible to achieve in real life. Furthermore, they believed that technology and its accompanying formal structure could be adapted and managed by the infrastructural processes of Indian culture. Their parents and grandparents believed and practised exactly such a compromise and were dubbed hypocrites.

The parental generation had attempted to reconcile differences in the social and formal systems. They realised very soon that the processes in the two systems were not complementary. They managed the discrepencies by making a distinction between, and developing differential behavioural patterns for, the two systems. The next generation demanded consistency between formal and social processes. Their stock phrase to justify this was that 'merit, reason and rationality' in formal systems must prevail. Action must be geared to the logic of the situation and not to the outdated idealism of the social and cultural systems. They held on to a belief which did not permit convergence between the formal and the social systems. In fact, in family settings they wished to perpetuate the tradition of the 1920s and 1930s, and that too with a sense of expediency rather than conviction. This finely differentiated and piecemeal integration of technology and economic thought led to imbalances in social living. It was obvious that the active historical process of Indian culture which had borrowed from alien cultures and assimilated them into the overall Indian agrarian ethos had collapsed over time. This led to individuals adopting the manifest forms of an alien society in the name of functionalism and progress. However, the relevant cultural processes and creation of infrastructures to support and maintain such manifest forms as were viable for that society were ignored, sometimes deliberately. Arbitrary imposition of such forms over the cultural processes of India made the structures more rigid. This created forces for further fragmentation, disintegration and lack of convergence in the processes of Indian identity formation and social living.

The data suggested that the Indian identity, as reflected by these individuals, was caught in a peculiar dilemma, not wholly of their own making. The dilemma had arisen and continues to exist because the educational system, the reformists and the mass media have over-emphasised and overexposed the negative side of Indian society and culture, underplaying its positive aspects. At the same time, exactly the opposite is done for Western society and culture.

The conflict of location of identity in lifestyle was deeply rooted in the psyche of the individual. It was the very primary struggle of 'being' and 'becoming' of an Indian in a culture of transience and in a society which was at the crossroads, where two or perhaps more sets of values and world-views competed with each other. The conflict of location of identity and lifestyle was only a manifestation of the choice of becoming for these individuals. In solitude they explored the issues of 'being'. They confronted questions such as: What kind of person should one be? How can he become the person he wishes to be? Who should he be, rather than what should he be in terms of social status, was more relevant. Here too, their choice was caught in a double bind with two opposing sets of values. The first set of choices was in conformity with the traditional ethos and the achievement of socially desirable goals. The second was personal satisfaction relating to changing times. This demanded active efforts to change the quality and boundaries of relationships around oneself and others. Both options left the individual with residual negative feelings.

The individuals who chose to conform considered the system outside them as powerful and themselves as helpless. Their feelings for the family, need to relate to colleagues, and fear of outside criticism prevented them from trying to reappraise or redefine relationships. They resented the pressures to conform and would have liked to assert and increase their area of autonomy. In effect they wished to make a choice. However, in the ultimate analysis, many of these individuals conformed but developed various strategies to cope with the internal conflict. A few of these are elaborated below.

What Parents and Others do Not Know does Not Hurt Them

These individuals conformed in the presence of social and formal authority to accepted codes of behaviour. In the absence of such an authority, they would indulge in traditionally forbidden activities. These might include drinking, for it was the 'in thing', wearing clothes which were in fashion, experiments with drugs to portray bravado, and presenting a 'mod' and a hep image among peers. They also spent a great deal of time discussing and expressing opinions, generally critical, of the system around them, its culture and society.

The System Should be Beaten at its Own Game

Some individuals, while apparently following all social rules in public, cut corners, took advantage of loopholes, took liberties, and gave excuses such as overwork, sickness, or personal distress to push the system to accept their deviations. Basically, they tried to manipulate the system. Occasionally some took to passive or even active subversion of the system. Deep down their scepticism and cynicism regarding the system were quite apparent. Some of them even claimed to have been sympathisers of the Naxal and other such movements in the past. Their best game was to catch out the system in its contradictions and then push to make it yield to their demands.

The System Should be Suffered

Very few individuals chose to suffer. They conformed, but demanded fairness and justice. They constantly grumbled whenever they found individuals who beat the system at its own game, getting away with it. However, they could not bring themselves to connive, manipulate or defy. Individuals who adopted any of the above strategies could not accept the fact that in their roles they were required to make efforts to modify the system. They felt that they did not have such authority. According to them, such authority lay only with the leaders of the system.

Those who opted for personal satisfaction also adopted certain axioms.

Do Not Act, but Prepare Grounds for Change

Individuals adopting this strategy talked of creating enough social awareness by voicing criticism so that the authorities could be made cognisant of current issues and problems and could, therefore, act in consonance with the ground reality. Some among them talked of the evolutionary credo. These were the individuals who appointed themselves as spokesmen in a crisis. They attempted to achieve a just and fair reformulation of norms and manage the interaction between the administration and members of the system.

These individuals believed that the system may have rules and regulations but they should not be regarded as absolute. Deviant action within the system must be evaluated situationally. It should be treated as an emergent situation and application of the rules should be mediated by consideration and discussion.

Their argument was: 'Yes, the particular individual has committed a punishable act. A reduced punishment should be given because of extenuating circumstances. Furthermore, these extenuating circumstances are becoming more and more common. The system should realise this and formulate new norms and policies.' The emphasis here was to make the system act while they provided stimuli for evaluation and change. However, many of their attempts sought change for the benefit of only their own group and not for the entire system.

Act with Conviction

Some held strong convictions and acted in accordance with them. In many spheres they had no problems interacting with the system and went along with it. In other areas these individuals acted in accordance with their convictions. They did not try to show off their independence, nor did they set themselves up as models. They were willing to be criticised or isolated for their action. Such individuals did earn a kind of respect in their community. Despite the fact that they were not very popular socially, they were very often sought out by people in distress for advice. These individuals assumed autonomy in choosing their values and actions, and paid its price. However, they did very little to influence the system directly; they supported it where they agreed with it.

The Self-oriented

Some individuals very provocatively designed their own modes of being. In fact, they mocked the concept of belonging to a system. They believed that one should extract the maximum possible satisfaction in life. Expediency was then their guiding principle.

In all these six stances and strategies, there was one common factor—all these individuals were concerned with creating some space for themselves in the system. Primarily, they believed that the system was not their direct responsibility. They were not concerned with its replenishment. Its integrity, wellbeing, direction and evolution were relegated to the authority, or to a diffused collectivity. None of these stances helped the individuals to develop a healthy sense of self-esteem. Some feelings of shame, guilt, mockery, disdain, and bitterness were always present.

These stances also reflected the individuals' relationships with the authorities of both social and formal work systems. Most of them internalised a historical perspective of society while relating to authority. They had either conformed, rebelled or defied authority. In many systems individuals exiled themselves from their families and organisations when they could not cope with the social and formal authorities. Similar patterns of the individuals' relationships with collectivity have emerged.

A critical glance at the historical process and the Indian ethos of the individual–system relatedness yields the following conclusion.

For centuries, following the Ramayana era, Indian collectivities have yielded to the hegemonic system outside themselves. The system has remained anchored in the will to power of the few dominant roles which define, run and control it. Within this pattern, leaders like Buddha have come forward to transcend the traditional system in order to gain greater space for the individual and give him back his self-authority. Yet time and again, such efforts have been nullified because the collectivity has chosen to perform prescribed roles. The tradition of being a representative in the system has always been left to a chosen few. We asked the individuals what role they proposed for themselves. The responses were very diverse.

System-centred

Some individuals argued in favour of supporting the stability and continuity of the system. According to them the privilege of redefining the values and the tasks of both social and formal systems must remain in the hands of a few legitimate individuals. It was clear that they might perpetuate vested interests in the name of stability and continuity and allow the system to remain rooted. Emerging realities were not realistically appraised but the individual considered this as a normal risk of belonging to a system.

Most individuals did not claim any responsibility for belonging to a system. Their idea of system and belonging to it was rooted in their experience which considered relatedness as being determined by birth, and hence involuntary. They treated their relationship to the system as a matter of accident or chance. The authority was always *outside* unless it was inherited through succession. According to them all systems must be centred around the authority of one person, a patriarch, or a charismatic leader, who controlled all resources of wellbeing and security. His governance of the system by patronage was critical for wellbeing. Accordingly, they had to remain dependent on the system and its significant leader even if he acted as if he were oblivious to the need for renewal and redefinition. They adopted a passive attitude or acted only through the formal structures.

We confronted this set of individuals with the fact that their orientation to the system reified them, so that they were always being 'acted upon'. They could only react and not respond. Furthermore, they were dependent on and worked for others. Theirs was the culture of passive obedience. In effect, they were denying themselves the chance of being resources for the replenishment and development of society. Further questioning led some of these individuals to admit that they were hoping a charismatic personality would appear to help them mobilise for a cause and provide a rallying point to bring about a change. They were hopeful despite the fact that the history of India is dotted with charismatic leaders, establishing cults and adding to the content of social thought, but without making any significant changes. This hope for change through a charismatic leader reflects a deeply imprinted mode of involuntary membership of society as well as a lack of commitment to initiate change.

Ideology-centred

Another set of individuals chose to align themselves with a heterogeneous blend of ideologies which included the thought of Marx, Lenin, Sartre, existentialists, Ayn Rand and other similar popular thinkers. A dialogue with them revealed their determination to manipulate the system. Their choice of an ideological system was cognitive and, not surprisingly, it lacked conviction for action. Their day-to-day behaviour suggested that their ideological stand was only a mask to hide either behaviour based on impulse or self-centred motivation.

Their thinking was characterised by stock words like *people*, *poverty*, *equality*, *fairness*, *human needs*, etc., and was confined to quotations and slogans from writers. They had made no serious attempts to question the assumptions of their ideology in relation to the process of living. Masked behind diverse streams of thought represented by Marx, Sartre and Ayn Rand, these individuals carried a sense of futility. For them the Indian system was beyond any significant improvement. It must be destroyed and replaced, ignored and flouted, or finally, disowned through migration. The only other choice which they shared with the first set of system-centred individuals was to acquire power, and through it control the system and find their freedom. Behind all these ideological stands, their fundamental belief was 'might is right'. Essentially, they were critical, isolationist and manipulative in their approach. They too had very little concern for the replenishment of the system.

Competence-centred

Some individuals chose to disengage themselves from any idea of belonging to the system. They anchored their identity in their professional, technological and analytical competence and said, 'We will go where we can use this competence. If India cannot use it we will go abroad. Our world is where we can optimise returns for our competence.' Their logic was tenuous. They had done their duty and society had spent resources to do its duty. They scoffed at the concept of commitment and replenishment. When pushed further, they said that the corrupt and hypocritical leadership operative in the society today could expect no claims on their loyalty.

In the ultimate analysis, these individuals believed in the *survival of the fittest*. They also saw themselves as superstars who came on the stage only to perform. The responsibility for backstage management, and the rest of the infrastructure for their performance, lay with others. They were not willing to invest any energy or effort in building, sustaining or improving the system.

Rooted in the idea of their own competence, their actions and relationships were merely instrumental and manipulative. Rather than face their own attitudes squarely, these people skilfully avoided the question of their relevance to Indian society and, as already mentioned, raised the issue of hypocrisy and corruption. In our understanding, none of them displayed any commitment to act towards improving the system. When confronted with this issue, they invariably stated, 'I do not have to stick my neck out. The system is too large for me to do anything about it.' To modify and change the nature of relatedness in their own immediate life space seemed a mammoth task and often led to withdrawal and anger.

Many of these individuals realised that they were only giving in to power. When demanded by their superiors to act in a certain way they modified their behaviour. All adaptation was thus made under compulsion or through enforcement. Even when they modified their behaviour, their cognitive structure remained rooted in the traditional model. The individual did not invest to improve the system, but continued to exploit it. In short, these individuals reflected the dynamics of 'disaffiliation' based on a strategy of survival and thrived in the martyr mode. The system was outside them, and it had to be treated as a constraint. It had to be appeased, placated, or manipulated for their benefit.

Trishanku[2]

There were some individuals who did not fall into these three categories. They were really the Trishankus. It was difficult for them to disengage themselves from the primary system and involuntary relationships. They shared many ideas with the ideology-centred and the competence-centred groups. However, their value system differed, comprising service to parents, denial of self for the sake of the family, and other traditional norms. Though they perceived their aspirations and life space as legitimate, they felt helpless and constrained. Attempts to break away from these normative boundaries and act in congruence with their intellectual capabilities generated guilt. These individuals did not grumble and often displayed a social attitude which persisted for about a decade after they had completed their education.

These four type of attitudes and orientations reflected the inability of the individuals to act at, upon or against the system. They were unable to take the responsibility for modifying systems or relation-

[2] Trishanku was one of the forefathers of Rama who wished to go to heaven as a living being and not after death. He did not deploy any effort to do this and sought the intervention of sage Vishwamitra who lent him the reward of his penance. Trishanku went to heaven, but was pushed back by the gods for illegitimate entry based on proxy efforts. As he was falling down to the earth Vishwamitra once again intervened and made him stand in the sky where he hangs till this day.

ships to realign them with the existential flow of the times. This inability seems to be a product of the culture of transience where the collectivity of society stands at crossroads. The poignancy of the individuals' experiences suggested that the Indian identity and the collective Indian psyche remain deeply buried in the cultural processes of the agrarian ethos.

As we searched for answers to these conflicts of the Indian collectivity and the resulting dilemmas, it struck us that these individuals singly and collectively were consciously or unconsciously enacting in modern times the psycho-cultural drama of Indian mythology (Parikh and Garg 1989). In the dominant mythology and culture the system has been, time and again, controlled by an absolutist and autocratic ruler, enveloping social existence and the individual's relationship to it in indignity and dehumanisation. The then citizen of India only cried for help—*trahimam, trahimam* (save me, save me). The myths relating to the ten avatars of Vishnu suggest that individuals and the collectivity disown their self-authority (autonomy) and responsibility in encountering centralised power. In doing so, they choose one of the four strategies discussed above, as do the individuals of today. Mother Earth then appealed for redemption. She presented herself in the court of Lord Vishnu and prayed to him to descend to earth to destroy the individual who represented the evil of power. He did so, but without changing the primary processes of society which led to gradual dehumanisation. It was thus the individual who got killed, while the processes remained unchanged. Hence in India's history of man's relationship with society the psychodrama repeats itself. The Indian collectivity seems to await the arrival of a charismatic identity who, as if by magic, will restore congruence between the individual and the society to live in peace forever after.

Apart from these core attitudes towards the self-and-system relationship, these individuals had a great deal of creative potential. They were capable affirming themselves in action. It is tragic that such competence, creativity and potential are being held in abeyance due to the culturally determined nature of the individual's relationship with the system.

The recurrence of the same process and the roles chosen by the Indian collectivity today raises the question of the origin and dynamics of the relationship, and the orientation of the collectivity towards the system. Is the current generation merely replaying the cultural psychodrama as suggested in the myth and reflected in the cultural lore? Why does the pattern persist over centuries? Are these active or passive responses reinforced by some of our institutions such as family, education and community? If so, which institutions? How do they operate to create such a strong hold on the psyche of the individuals and collectivity?

The cultural history of India suggests two different, but distinct, strands whereby congruence and convergence in the relationship of the individual with the system were attempted in the past.

Individual-centered action

From the hoary past of ancient India to the beginnings of known history there have been many individuals who stand out. To name a few, they are Mahavir, Buddha, Sakubai, Gorakhnath, Nanak, Kabir, Raidas, Ram Mohan Roy, Dayanand Saraswati and Gandhi. There have been a large number of individuals in each era who, by the courage of their convictions, redefined values, reformulated social ethics, and initiated new world-views. Unfortunately, the Indian collectivity in its response through faith has always made them into cult figures, then deified and institutionalised them. Their impact was thus limited to a certain set of individuals whom they could influence and modify. But the processes of cultural collectivity which define the relationship between the individual and the system remained unchanged. The impetus for individual action has always been tied to the faith of those outstanding individuals and their assumptions about man's existence, his relation to a cosmic order, and purpose or being.

Collectivity-centered

From India's ancient past till the period of Harshavardhan, the King of Kanauj, there existed a tradition of Dharma Yagna or Rajsuya Yagna. From the writings of scholars like Acharya Chatursen and others, it seems that this institution was designed to take stock of the projected and emergent social phenomenology of the times. It also aimed to document dysfunctionalities, deviations and departures from the coded ethics of social behaviour. This institution was nurtured by the king who invited intellectuals from various disciplines and different regions to meet for a dialogue. They were to examine dislocations in the structural and interpersonal codings and the resultant structures and modes of relationship and transactions in society. The objective was to arrive at a consensus to recommend structural modifications, create new codes of conduct in society, integrate new groups of people arriving from outside the country, and recalibrate and realign the individual's relation with the social structure so that a new coherence and convergence could emerge in the diversity that was India. This institution kept alive the reverence for the social design which could sustain change and contain diversity in order to preserve a collective cultural identity. This collective action was not based on faith and conviction alone but was anchored in the rationality of the interdependence of social living and inevitable change.

We continued our search for the primary sources of Indian identity and were struck by the fact that Indian culture does not allow for the resolution of primary ambivalences that all human beings in their processes of socialisation experience with significant roles and the system. The individual is left with the burden of making decisions at every step, like Arjun, between what is 'kartavya' and what is 'akartavya'.

Our experience with our groups shows that individuals continue to hold, simultaneously, opposing and often contradictory feelings about themselves, others and the system. The positive feelings of self-worth and wellbeing, and the capacity for constructive action and assertion exist on a substratum of self-hate, self-defeat, impulse to destroy and passive aggression. Doubts, anxieties and preoccupations keep pace with agreements and the capacity to take hard decisions. The possibility of hoping, and visions of a worthwhile future fall victim to such ambivalences. At the macro societal level, the current national, social and political scene displays the same configuration which is confirmed by the media. Individuals and the collectivity are intellectually aware and in fact know what is wrong and what are the choices for action. Emotively, they wish to initiate action but end up conforming to the system as it is. As such, no action on change occurs. However, the system which is actually capable of such actions is preoccupied with the complexity of relationships and resources, and thus fails to fulfil the expectations of the collectivity. This process leaves the collectivity helpless and angry and adds to the existing dysfunctionalities, chaos and disorder.

The unresolved expectations have their repercussions in the relationships with siblings, peers and colleagues. The Indian collectivity is incapable of managing differences at the peer level. Most agreements end up as pseudo-compromises which break down in action. There seems to be an inability to clinch and consolidate any success. This mode likens the Indian identity to the myth of Sisyphus.

The exploration of the Indian identity, its continuities and consistencies over centuries, and the emergent discontinuities and sense of discreteness in the current generation, creating a culture of transience, are the themes of this book. This culture of transience is experienced by the men and women of today. We attempted to understand and articulate the emotive map as experienced by this generation, together with the cognitive map of Indian society available to it through education and other sources. Both the emotive and cognitive maps of individuals and the emergent structure and processes of the collectivity need to be understood if any constructive alternative for evolving a new society is to be found.

This book presents our attempts to understand and describe the current cognitive map of Indian society in the context of this cultural transition and the resulting transience. In order to delineate this map, we recorded the experiences of individuals with social institutions, the family, educational institutions, their neighbourhood, peer group and work organisations. These explorations revealed an interesting mesh of two kinds of historical processes and data: (*i*) the process of the individual's own growth and development, and (*ii*) the process of the emergence of the transient culture constituted by the interaction between a traditional, agrarian society, and a modern, industrial society.

Most of the individuals we studied, while talking of their own processes of being and becoming, tended to reveal a great deal of what they had internalised about the nature of society from their grandparents, parents and other people. This book attempts to integrate the experiential data reflecting the nature of Indian society, its structure, processes, values and ethos. This society and integration with it, as lived and experienced, perceived and given meaning to by these men and women during their process of growth, is our theme—one fraught with pathos.

While listening to the shared experiences we became acutely aware of the emotional overtones involved in the dilemma of choice, and the stands taken towards the system described earlier. These emotional overtones are significant in understanding inconsistencies in the interpretations of society given by these individuals. Their universe was an interplay of two cultures constituting the setting of their growth. We became aware that these individuals were growing up in a culture of transience. Much of the discussions in the research literature has centred around either the polarised conflicts between elements of the two cultures—Indian and Western—or the evolutionary/Darwinian model of analysis. We have attempted in this book to bring out the constellation of the culture of transience.

The culture of transience as it emerged from the descriptions of these individuals sometimes appeared like an assemblage. In some instances it integrated into a whole. At other times it appeared as a mere juxtaposition of elements, or nothing but an amorphous, turbulent heap of rubble, a quagmire, an unstable quicksand, ever changing in its manifestations. Any new input was sucked into this unstable transience. One neither knew nor could predict the direction and quality of the next state of the culture of transience. In our attempts to make sense of this pattern we had to face our own cognitive and emotive maps of Indian society. This made our task very difficult. Every encounter for us was full of pain and turmoil. At times it made us angry and at others very small and invalidated. In order to integrate the contradictory data of individuals into a coherent pattern it is necessary to site our own position. This is to try and discover the emotive and cognitive maps of the society, collectivity, relationships and roles. We discovered a perspective in ourselves which we had neither reflected upon nor articulated. We have called it the ignored perspective.

The Discovery of an Ignored Perspective

Since 1972 we have sat and listened to the rendition of the private and personal experiences of many people. We have been moved and touched. We have been taken aback, full of admiration. On many a occasion we have been most indignant as we have through their pain and torment. Sometimes, for days we have waited anxiously, watching a young man or woman struggle desperately with his or her transactional and existential theme, in an attempt to come to terms with himself or herself. And all the while, we have been absorbing their experiences and forming images of the context, the societal field, and the socio-cultural setting of their growth. Gradually, the vague and diffused structure of the context, the societal field, and the socio-cultural setting began to crystallise. Many questions arose: What is this country of which we are a part? What is our heritage? What is our culture? What are the context, values and the ethos that we have grown up with? These and other questions always surfaced as we relived with the men and women their dilemmas, joys, anger, disappointments and the whole series of experiences with relationships while growing up.

Our involvement with them brought back the awareness that behind the manifest world of joy, excitement, education, achievements, social events, and youthful escapades, there lies a world of turmoil, populated by anxiety, doubt, guilt, and an ever shifting sense of adequacy, security and wellbeing. Fear of acceptance, affirmation and approval, pangs of loneliness and unwantedness punctuated with moments of thrill, excitement, spontaneity, bursts of assurance. As we listened we could not but join in with recollection of our own experiences of being and becoming. We have found ourselves in resonance with the whole spectrum of feelings they voiced.

However, when we sat back and reflected on the private and personal world of these men and women and our own experiences, we were disturbed as we realised that the meaning the current generation give to their context of growth, societal field, and the socio-cultural setting was at variance with ours. The difference intrigued us, and we started to wonder about the change that has taken place within the last fifty years in Indian culture and society. Our cognitive map and theirs had many similar elements. But somehow, the whole configuration, the coordinates and underlying meanings and processes had changed. The landmarks appeared familiar but the contours were completely strange. To us, the Indian social system was accepted as a closed field. Very few of us dreamt or aspired to live abroad. For some of us education and for that matter education abroad was only a dream. Our commitment to live in the system was high. The foreign government was a cause to fight against. It made us have a national identity anchored in a positive cultural heritage of the past. It also forced us to identify the strength of the system and we were acutely aware that the reformation and restoration of society was a powerful theme in our lives. We had a sense of involvement with the social system. We accepted social authority as legitimate, and binding on us. Alternatively, we would attempt to modify its role in our lives. For many of us family remained the sheet-anchor and we were willing to treat ourselves as its extension. Mutual obligations were recognised as part of role relationships. The network of familial relationships was a strength, supporting us in our achievements and the fulfilment of our aspirations.

The social system was not only a context of life, we were committed to it, and it often set our direction and goals. We accepted the responsibility for replenishment of the system, even though in action we rarely did it. Work was a means for fulfilling commitments to the social system. Norms of society, though resented, were largely preserved. Departing from them was an act of conviction or impulse, rather than one of reason. It was done with the purpose of making a bid for change. We rarely departed from norms just because we needed to declare our independence.

All these considerations and attitudes preserved in us some sense of belonging. We were proud to be Indians. Exposure to the Western way of living and thought was kept outside the home. Cognitive acculturation in the Western ethos did not lead to our disengagement from the Indian ethos. While in school we read Shakespeare and Greek myths and legends, at home we were continuously exposed to the Puranas, the Indian languages, and occasionally to Sanskrit literature. We were, on the whole, content with our own social system.

To the current generation the social system is in a disarray, littered with outdated but compulsive residues of stagnant forms. The young see the family as a feudal system. Their exposure to Western thought and lifestyle has created new a framework of independence and skill-based occupations. In the comparative frame which often creates negative images the Indian social system is seen as wanting. It also creates a strong reluctance to examine the realities of the Indian social system. As a consequence the current generation does not have a sense of belonging anywhere, except to the world of ideas and technical skills. The written word from the media becomes their benchmark for defining reality. Except for achievement in terms of high performance in education, and intellectual efficiency, there is very little personal challenge for integration of their identity.

Unlike what it was in our times, the Indian scene is no longer populated by a galaxy of giants in every field of life. To name but three giants of our times—Tagore in literature, Gandhi in politics, and C.V. Raman in science. Besides the national figures, almost all the states of India could name many in literatures, politics, science and other fields. They provided role models for identity integration. Now the challenge of reformation is over and there are no heroes.

Today's heroes, created by the media, are continually being replaced by new fashions, fads and marketing strategies. There are slogans only for development and forging a prosperous India instead of triggering processes for integration of identity. This pushes the current generation into the soft option of material acquisition. The family no more acts as the sheet-anchor of belonging. Overexposure to the international scene highlighting the lacunae of Indian life undermines the sense of pride in being Indian. The current generation has become disengaged from their life experience and has learned to live by well-formulated opinions. They have become defensive, argumentative, arrogant and even hostile.

Discussions of our experiences of growing up and India's past and their present experiences created a comparative frame, entrenchments and resentment. This created many a tense situation. We were told that we were chauvinistic and sentimental. The young men and women used Ayn Rand to justify their stance. They tried very hard to educate us in the philosophy of individualism which to them implied scepticism of historical continuity. They saw society as only taking freedom away. Their interpretation of Ayn Rand was very often combined with the Marxist interpretation of history, and an anarchist view of society as being only coercive. Of course, these were some of the extreme stands. The dominant stance was that of committed scepticism. The general feeling was of being let down by the system.

Eventually, a stage came when individuals confronted us, and each other with the question, 'So, what next? We know what we are doing, what we are letting others do to us. We also know that we have to do something—what should we do?' Invariably, the individuals turned to seek behavioural solutions for identity dilemmas. It became a question of learning to manage and manipulate the feeling and action system of the self. Inevitably, the behavioural solutions we advanced were opposed. We realised that our solutions were anchored in our own past realities and were of little relevance to the current generation.

At this juncture we realised the present generation's desperate struggle to come to terms with what they believed to be their values, their culture and their ethos because it is in these that the coherence and integration of the self and role identity, can be anchored. Without such anchoring, individuals drift and attempt to adapt themselves to the environment around them. Their need to come to terms with themselves and the system was tied to their search for convictions to live by. They had admitted their a strong core of scepticism. It was with some difficulty that they recognised that their propensity to conform, whether to the orthodoxy of action and thought or to modernism, was merely acting without convictions. They also recognised how this lack left them full of self-doubts and a harrowing sense of inadequacy, even after they had repeatedly attained socially desirable goals. They admitted their anger, resentment, and even hatred for the system to which they belonged.

As we attempted to trace a pattern we became aware that the processes of Indian society and culture as experienced by the young men and women might very well be completely at variance with those established by the socio-economic, political, historical, and religio-philosophical analyses of academicians. We had already established that our own experiences of Indian society were different from those of the young men and women with whom we held the dialogue. In the initial stages, we were using our own reality as a benchmark to compare and evaluate the current generation's experience of reality. Later, we realised this distorted any attempt to understand current reality. Conflicting interpretations of reality are natural, and have to be accepted and understood.

How then does one deal with this variance in order to make it the basis for proactive action? In the process of seeking answers to this question, we recognised the need for developing a psycho-cultural perspective within which the socio-political and other phenomena can converge to provide meaning and relevance. Without such a perspective, analysis of any of the above phenomena leads to fragmented diagnosis. It also does not release any synergy for planning enduring action. At best, it provides choices for short-term 'patch work' action aimed at reformation rather than regeneration. Resistance to such action is natural. This is what has been happening in the last fifty years in Indian society. We believe that for a stagnant, conflict-ridden society and culture on to take the first step on the road to regeneration, the discovery of the psycho-cultural perspective is both central and essential. This has helped us recognise the need to examine the heritage and culture of India. Today's generation is looking for a contextual perspective for validating their being and becoming. Focussing only on feelings and actions is insufficient for resolving concerns and sustaining developmental action.

To sum up, we are aware that some parts of our analysis and construction of Indian ethos and social structure, as it is experienced by the current generation, may not fit the academically and traditionally accepted perspective. However, we believe that reality is that which is experienced, felt and responded to and is never only that which is ideationally and logically established by academic discipline. Our intention is not to belittle the objective and scientific interpretations of Indian ethos and society, which have their own place. But rigorous and qualitative research in portraying and establishing the living reality of society through the psycho-cultural perspective is also a valid research. In our experience such qualitative research is essential if change in culture and society is to be understood or planned. We also became aware that like an individual a society also has a cultural identity. The universe of cultural identity of society and individual is anchored in the assumptions and structure of the society, which we set ourselves the task of exploring. We realised that the task is rather difficult. The available scholarly research presents several constructions of Indian society. They vary drastically from each other. Consequently, we decided that we would fall back on our personal explorations as well as those of our participants to discover some basic assumptions and frameworks of Indian society. This experiential reconstruction would be closer to the reality of our own transactions than any one constructed by scholarly models. In the next chapter, we present an account of how the men and women experienced the Indian reality in order to acquire a semblance of identity.

Two

Reliving Indian Reality: Emotive Maps from the Agrarian Ethos

A Content Perspective from Outside

Each of us carries within us a unique pattern of our heritage. From the cultural, social, historical and family sagas we internalise and structure our experience. Helping us to coordinate these patterns into a map, interaction with adolescents, youth, managers, housewives and men and women from diverse sectors of society flooded us with data and perspectives. The data had many strands and beginnings. Sometimes the quantity of data led to confusion and disarray. When individuals walked down memory lanes recalling conversations, experiences and interactions with their parents, it reminded us of our own childhood memories of flying kites. Climbing on to the rooftops; letting the kite fly high in the sky; getting it cut and losing it; being anxious to retrieve some of the string; pulling it as fast as possible, only to drop it in loops and then sitting down to rewind it onto the *firki*. In the process one encountered multiple knots, twists and turns, entanglements and generally got lost in trying to trace the continuity of the strand.

Our attempts to explore the Indian reality as experienced by a diverse set of individuals, was like holding a whole set of strands with many knots. In this attempt the first set of knots led us to ask the following basic questions: What is this country Bharat? What is this society India? Who are the people who call themselves Indian? The answers to these questions were as varied as the data.

India is an experience—kaleidoscopic, profound, cataclysmic. Few respond to this experience with indifference. The country, its society, its people, its environment and its ethos evoke both praise—varying from mild to rapturous—and disgust, varying from contempt to condemnation. To some, India is a feeling; has a splendid and glorious past; and is a country steeped in concern for the spiritual wellbeing of man. It is a model of hierarchical society, and has a tradition of searching for harmony in the midst of turmoil. Visitors and travellers tend to look upon India as a wise and experienced teacher, one who after four thousand years of experience stands as a symbol for the indomitable spirit of man. They come on pilgrimage as seekers of the human spirit. Some lose themselves in their search and ignore the harsh social reality. Others get caught up in the harshness and give up their search.

To yet others, India was, and still is, an exciting dream—the Land of the Golden Bird, the land of prosperity and beauty, of riches in wealth and in people, and of dignified and elegant men and women. A land of immense variety, evoking dreams of heavenly sensuousness and material gain. They came to India to plunder, to subjugate and to exploit. And to still others, India with its tropical climate was, and is, a country of vast and varied panoramas, a slow and slumbrous countryside traversed by rivers, topped by mountains, and filled with colourful birds and people. To them India was a fare to be delicately savoured and dwelt upon in slow sweetness. Their dreams evoked images of swaying bodies with feet dancing across the land. They heard the haunting sounds of ponderous chants and reverberations of gongs from temples, the call to the faithful from the minarets of the mosque, juxtaposed with the chimes of cowbells, the chorus of bird-songs and the fluttering of wings as thousands of birds settled down for the night. They talked of a whole way of life in perfect tune and harmony with nature. They came to an exotic and picturesque land as spectators and wanderers.

From all these images it seems that India has been perceived as a source of fulfilment—sensuous, spiritual or material. Many of these images survived the early stages of the arrival of the Europeans. However, since the 1920s another set of experiences of India as a country, society and people began to emerge. They were products of two processes: one, the political reorganisation of the infrastructure of society which had temporarily provided India some strength and cultural cohesion in the midst of confusion and uncertainties; two, the writings of some scholars and administrators painting India in terms of its dysfunctionalities. According to them, India was a land of perpetual famine, rampant sickness and dire poverty. It was full of eyesores, beggars, cripples and lepers. 'India' became synonymous with squalor. It was, in addition, peopled by a few ostentatiously rich maharajas, thugs and criminals, dishonest businessmen, corrupt officials; all in all, the epitome of a lack of human decency. It was seen as a country of illiterate and superstitious people, steeped in the prejudices of caste and religion, perpetuating discrimination, deprivation and exploitation.

It became a country of people who clung to their traditions and refused to change; who, besides being fatalistic and averse to change, refused to act for themselves; who cried like helpless children pleading to be saved, clamouring for foreign control. It became the country described in the writings of the Katherine Mayos, Nirad Chaudhuris and V.S. Naipauls, which saw nothing but distortions of human existence in Indian society.

Between these two extremes—uplifting rapture and scathing condemnation—we suspect that the reality of India has been lost. What is this reality that bewitches and bewilders? What is the reality of a culture and ethos which holds together such diverse elements? What is the reality of a people who have survived over centuries, undaunted, unashamed, exploited, invaded and yet treading their own path and desperately clinging to something intangible? Perhaps an identity yet to be understood. The current generation has unconsciously borne the burden of these images of an Indian reality. They were acutely aware of India's vast variety and diversity, and that most images were reflections of reality, but they could not identify the strengths which held them together. They were divided among themselves in their reactions. Some thought in positive terms, others in negative. To accept the simultaneity of existing contradictions was mind-boggling. They were unable to grapple with this multiplicity. When pushed to extremes, they either adopted a logic which led them to scepticism, or a fatalistic acceptance. This led to disengagement from the system. It is tragic that in their long years in educational institutions, they remained ignorant of the vast literature about India as a culture, its dynamics and its identity. In order to initiate the search for this unknown identity, we brought together our own understanding of the literature of India.

Systematic research towards understanding India's culture, its dynamics and its identity only began during the last two hundred years.

Attempts prior to this were largely descriptive and subjective accounts of individuals who tried to make sense of their own perceptions. They were not accounts of researchers following a discipline. Their primary data was based on the manifest kaleidoscopic social behaviour and physical settings to which they were exposed. To all of this they tried to give meaning by collecting folklore or based on their own perspectives of events. This tradition continues till today in quite a few writings. Records of systematic research began with a rather narrow stream, influenced by the European ideologies of the nation state. By the 1930s this stream had broadened to a vast river. The following major trends in this research can be identified.

Macro-analysis of the Ethos and Culture

This is the tradition of Max Mueller, S. Radhakrishnan, A. Basham, Radhakamal Mukherjee, Jawaharlal Nehru, Rahul Sanskritayan, V.D. Agarwal and many others, some of whom wrote in one or other of the diverse Indian languages, others in English. These writers delve into India's past, reconstructing it in terms of its spiritual ethos, metaphysical values, life orientations, and the historical continuity of its social organisations and institutional designs. They provide an elaborate appraisal of the past in its various aspects presenting a positive picture of India. They confirm the idealism of the Indian ethos, and provide a succinct backdrop for the evaluation of its relevance today. However, they cannot help in understanding the complexities of the contemporary Indian reality.

Macro-analysis of the Institutions

Twentieth century social scientists like Nirmal Kumar Bose, Goetz, Ghuriye and Irawati Karve began systematic studies of institutions such as family, caste and marriage. Disciples of these, themselves stalwarts, like M.N. Srinivas, I.P. Desai, S.C. Dube and others moved towards specialised studies of villages and communities, in keeping with the trend of post-war American social science.

This macro-institutional approach deals with the basic data of Indian life and systematises it to interpret the nature and dynamics of the Indian social reality. However, the approach is mainly concerned with academic issues of definition and classification. Occasionally, one can glimpse a comparative framework whose axes lie in broader social theories. These authors have done valuable work creating viable categories, while also examining social change. In addition, attempts have been made to assess the macro-dynamics of societal development, where goals are set in terms of a western consumer-oriented, industrial society.

We wondered how worthwhile this was. With the help of this tradition one can observe the social structures, described within the framework of a chosen ideology. The explanations can be used to redefine the problem of analysis/diagnosis. If experience is any indicator, the efforts in this direction have been mainly academic, have provided *post-facto* explanations for failure, and created divergent opinions about what should be done to change Indian society.

Microanalysis of individuals and unit communities

This is the third trend which began largely with the arrival of American social scientists after the Second World War. It runs parallel to the tradition of macro-analysis of the ethos, culture and institutions. The unit of study is the individual or community. Overtly, and often covertly, it is anchored in a comparative framework. It also establishes a tradition of drawing its perspective from Western scholars like Opler, Erickson, Shield and Carstair. They, having looked at a molehill, speculated about the nature of the mountain and claimed to have understood it. Much of this work tends to stand in the way of understanding Indian reality. These researchers seem to use a preformulated model and theoretical framework to explain society. Their implicitly evaluative stand is naturally related to their hypotheses, goals and directions. The overall tendency of Western scholars to interpret all secular social phenomena of the Third World in religious terms is also evident.

In our attempt to understand Indian reality and identity, we extensively explored these three trends. We tried to relate them to our work with the generation of today, as children of two cultures. The three trends were inadequate to understand our living reality, where massive inequalities relating to overpopulation, poverty, illiteracy, superstition, health and casteism generated shame and, as a consequence, hate for the self, system, society and culture. These feelings led to the urge to reform and change our society, setting up Western models as the ultimate goals. In trying to explore only the Indian roots of our identity, the literature of Indian authors reflected some coherence and congruence with the experienced Indian reality. It articulated the processes and dilemmas of growing up within the sociocultural context of tradition. The realities of living and feeling and the resultant themes which emerged in the lives of the fictional characters made some sense to the current generation.

The Indian reality reconstructed by the social scientists reflected the event–structure of life, but had very little resemblance to the experiential quality of reality. The individuals agreed with the analysis as far as it went but had many 'ands' and 'buts' to add to it. This divergence was both intriguing and enlightening. For us, it reestablished our notion regarding the sphinx-like quality of India—its inscrutability and multifariousness, a collage of diverse unmatched pieces which hung together but could be seen as individual fragments as well. It made us recognise the quality of the Indian scene which provides as many interpretations as there are people and perspectives. Each could, and many did claim, 'This is it, this is India.' However, as we processed most of these accounts against the backdrop of the personal experience of growth, of different sets of people in many personal growth programmes, we could only repeat, 'May be it is, but is it really?' Very often the sutra, '*neti*, *neti*' (not this, not this), came to us as we sat reflecting about Indian identity and reality.

Our attempts to explore the Indian reality and its linkage to the Indian identity left us in a dilemma. While the manifest contents of the spectator–commentator position could not be denied, it appeared that emotive biases prevailed. Interpretations fell into two groups—those which glorified Indian reality and those which denied and condemned it.

Our exploration of Indian writings and the scripture-based explanations of society made us happy and proud of being related to a glorious past. However, while our sense of mystery about the past increased, our understanding did not. What India and Indians were, according to the texts, seemed to have little experiential relevance to what Indian identity was. Against this sense of shame and guilt was set the Western, alternative comparative model, leaving only the option to accept evaluations of Indian identity which ran contrary to our experience.

Historical reconstruction, describing life as it was lived in certain selected periods, improved our understanding of the ways in which our forefathers approached life. We now know where all their 'shoulds' and 'oughts'—articulated values and beliefs—came from.

It also became clear that the parents of the current generation have been squarely caught in the moral dilemmas of a traditional agrarian society pitchforked into a modern industrial culture and society. This dilemma has generated incompatibility between their actions and beliefs, due to the conflict between role orientation and self-orientation, between feelings actually experienced, and between social identity and work identity. This understanding helped us and the participants resolve some of the emotional issues with our parents. But it did not provide us with a choice for action.

Our exploration of the writings centred round macro-analysis of institutions helped us to some extent. The micro-analysis of units, individuals and communities sensitised us to happenings outside, but gave almost no clues to link them to our own identity themes. Thus, the exploration of different versions of Indian reality made us eventually fall back on our own resources. However, the glimpse into the vast panorama of India's past and present renewed our dedication to look for what is India today and what it is that contributes to our being ourselves.

The *Sthita Pragya*: A Process Perspective from Within

In our search for anchors of identity, we realised that India, a surviving culture and civilisation, is a vast melting-pot. From the beginning of recorded history, ethnic groups with their unique political, socio-economic and technological forms and processes, beliefs and values, entered the subcontinent. Nowhere else in the world has such an admixture of ethnic identities and juxtaposition of lifestyles been woven into a dynamic living society. Each group preserved its ethnic identity and lifestyle. The United States of America in modern times can claim a similar privilege of being a vast melting-pot. However, there is a difference. The process of Indianisation is the unique characteristic of the melting-pot that is India. The process created an overarching psycho-cultural identity, and simultaneously preserved the specific identity of each group with its diversity of myths, social forms and behaviour. The diversity in language and religious rituals was retained through its unique social design. Indian society has not forged a standard or uniform set of behaviour patterns and life orientation, as has been done in the process of Americanisation in the United States of America.

The Processes of the Ethos

The notion of the Indian ethos as available from the *Vedas*, *Upanishads* and the *Brahmanas* is hierarchical in nature. It can be recast in a strong and rigid ritualistic religion and made to prescribe to a fixed world-view. It can be anchored in faith and the supernatural, like that which emerges from the Judaeo-Christian tradition. It can view life in a linear and sequential fashion by establishing cause-effect relationships between events. It can also be interpreted in a philosophical, secular, non-ritualistic mould. It can encourage movements as diverse as Buddhism and Jainism on the one hand, and as empirical as that of Charvak on the other.

The resilience of such an ethos has allowed Indian society to absorb many alien strands and weave them into a multicoloured fabric. This process has helped the Indian people to retain a sense of continuity, direction and meaning without feeling estranged or threatened by the loss of identity. This resilient quality has been an insurance against psychological and spiritual decay, which has been the fate of many ancient societies and civilisations. The ethos of those societies was bounded, contained and rigid. They could not sustain violent changes nor allow alternative world-views to emerge from era to era. Indian society retained its vitality and responsiveness throughout the many socio-economic and political changes it underwent. In essence, the Indian ethos has the quality of maya itself. The word maya does not mean illusion or illusionary phenomena; its root meaning is transience.

The Quality of Psycho-cultural Processes

The psycho-cultural processes of Indian society allowed individuals or groups of individuals to give differential meanings to their relatedness with the environment. The process also allowed space for the adaptation of transactions to the locales of social living, enabling the development of a wide variety of divergent forms, manifest behaviour and achieved harmony in the lifespace of people. But behind all the diversity of manifest forms and behaviour in transactions, a singular unity of meaning and identity was managed by an overall social design. The life space of most individuals was postulated in such terms as having the freedom to seek one's own realisation and salvation without impinging on the similar freedom of other individuals.

This ensured a sense of psychological security for the individual in the midst of social diffusion.

The Social Design

Indian society displayed a unique social design. It emphasised self-sufficiency, containment and operational autonomy for each village unit. The lifestyle of each village, a micro-unit, was allowed to be anchored in the immediate environment. No attempts were made to create a distance between the life needs and the means of meeting them. Each unit had its own social authority. Thus, all issues of an interpersonal nature could be dealt with within the self-contained unit. The basic principle involved in the social design was distributive social authority but not decentralisation of administrative power.

At the second level of social design these self-contained micro-units had been placed in a cohesive arrangement of affiliative relationships within the main structure. Each micro-unit had links with a set of other micro-units, and thus acquired region-bounded significance. Distributive (and not decentralised) social design allowed for integrative processes through a secondary structure of affiliative relationships on the one hand, and cultural integration through the nature and quality of ethos and psycho-cultural processes on the other.

Indian design created a social structure which restricted life within the bounded space of the micro-unit or within the regional locale. This generated tremendous pressure on the individual. It bound and contained him within the limited transactions prescribed by the social design. In spite of the resilience of the ethos and freedom for divergence at the manifest level, this pressure accumulated and created psychic tensions which were often expressed in deviant and pathological ways. Indian society in its social design created a model therapeutic community to manage this so that the core and substantive ethos, psychocultural identity and social design could remain largely operative. This therapeutic community revolved around festivals which provided a well-bounded socio-psychological space for expression of these tensions. Each community added to the main ritual of the festival a set of second-level activities, some of them almost licentious. These activities could be carried out in public without shame or fear of punishment only on these occasions.

The Value Assumptions

The value assumptions supporting the processes of Hinduisation and their four distinct characteristics, viz., Indian ethos, psycho-cultural processes, social design and social infrastructures, were manifold. However, they were derived from a single crystallised, cryptic sutra, *Aham Brahmasmi* (I am the Brahma). This means that I, the self, the unit, the microcosm am identical with the macrocosm, and am therefore the macrocosm. The unfolding of this sutra unravels the religio-philosphic ethos of much of India. The first level elaboration of the sutra can be stated as follows: Man is a part of the infinite and, as such, is infinite in nature. His purpose of being is to grapple with the experience of being infinite. The infinite, however, cannot experience its own infiniteness except through taking finite forms.

The consequences of the first-level elaboration is that life is a continuum and death is merely a punctuation mark, a comma. According to this proposition then, man, who is simultaneously a microcosm and a macrocosm, takes finite forms to experience himself through a series of life-cycles. Thus, a new proposition of the theory of reincarnation is stated, making man's life timeless and beyond the specificity of social phenomena. In order to explain the varied locations of microcosms in the world at any given time, a theory of karma is propounded. It explains beautifully both the variety and movement of the different microcosms through the life-cycle and the variations in the social hierarchy.

Thus this basic sutra became the cornerstone for designing social organisations, with its multi-level elaborations involving the theories of reincarnation, karma and transient social reality. It led to the following first-level boundary conditions in social design.

Boundary Conditions for Social Design

The first level boundary conditions had two components:

1. Each individual, no matter where he was born, would be in the process of unfolding his karma and going through the process of realising his infiniteness. His life boundaries and his identity were sacrosanct and inviolate.

2. The role of society would be to provide space to the individual in the social structure for unfolding his being. Life was a process to be lived, reflected upon and transcended.

These boundary conditions led to the development of an ideology of non-interference, non-aggression and stasis. As a result, society carved out for each ethnic group a space which allowed its membership to retain its identity and yet be linked to the fabric of larger society. The social design was meant to preserve the sanctity of the individual's lifestyle and orientation, and to reinforce it. For example, it is not unusual to come across in the same village four different castes of one profession, say weavers or fishermen, who do not intermarry or interdine. Within the same profession, technology, myths, folklore and also customs and rituals differ. Even definitions of such concepts as incest differ.

It is within this primary religio-philosophic framework that the Indian agrarian society was both heterogeneous, highly diversified and differentiated, and homogenous, unified and integrated. In times of social turmoil, the Brahmanical elite resystematised the social codes and recodified social linkages retaining the unit-level autonomy and regional self-sufficiency in socio-economic and political processes.

The process of reorganising social codes and linkages is analogous to the organisation development (OD) process of the modern era. Evidence of this process can be found in the 108 *Smritis* which were written over a period of a thousand years. All of them were inspired by the primary code of *Manu Smriti*, but reinterpret operational principles according to the time and locale. Acharya Chatursen points out that there was a practice for the dominant king of each era to hold a dharma *yagna or rajsuya yagna*, where scholars and propounders of social law were invited. Conferences were held, lasting over days, where after debate and discussion the social codes were restated. At the conference, emergent dysfunctionalities in the social order were taken up and modified. The *Smritis* are merely recordings of these decisions.

These ontogenetic corrections of social codes and reinterpretations of value assumptions into newer operational modes are also reflected in the tradition of commentaries on scriptures. These commentaries reflect the dynamic tradition of modifying the *shashyat* or the eternal assumption into a relevant form for the yuga, i.e., the current era.

These commentaries were largely experience-based restatements of the principles in keeping with the realities of the times. The process reflects the attempt of the ruling class to reaffirm the basic principles by reinterpreting and reoperationalising the ethos, according to its own needs. There has been no other culture where such systematic efforts to retain, for a privileged class, the basic sanctity of man's value structure, his basic freedom to work out his life, and his basic sense of self-worth as a human being.

In this tradition, the word 'value' has three distinct meanings, each represented by a different word. Each word locates 'value' in sequence from normative to phenomenological to existential levels. First is the notion of *mulya*, the normative behavioural values dealing with options. For example, Indian ethics recognises eight kinds of marriages, including *gandharva vivah* which is analogous to the informal living together of couples today. It also included rakshasa *vivah* which made allowances for forced marriages, rape and abduction; and *niyog* allowing a woman to marry temporarily to have a child by another man if the husband was not available or incapable for a long time.

The second word for value is *pratigya* or dharma. This value implied choice within the social context which an individual exercised in order to give meaning to his life. This set his boundaries in transactions and provided him with a clear-cut sense of continuity and security. Part of one's *pratigya* or dharma was shared by the community group, *kutumba parampara*, but the rest was called *vyakti* dharma thus allowing the individual to carry on the group socio-cultural heritage as well as enrich his individual heritage.

The third word for value is *aastha*; it is very close to the word ethos and refers to the existential level. *Aastha* encompassed a set of processes and propositions unifying normative and phenomenological values. 'Ahimsa *parmo* dharma' (non-aggression and non-violence) is one such *aastha*, even though the meaning of non-aggression and non-violence differed from era to era, from religion to religion and from community to community.

The political and economic structures of Indian society gave rise to the following boundary conditions.

Distribution of Resources

Preservation of identity, in which the total lifestyle was contained in a social space along with the technology, myths and world-view, can be illustrated by the various castes of weavers categorised on the basis of distinctions in professional orientations. This was reflected in differential looms, technology of craft, processing of raw material, patterns of weaving, choice of dye, design of the fabric, and even texture and pattern of the final product, the cloth, which was unique to each group of weavers. The differentiation inevitably led to a product market segmentation. If we examine this and many other examples, it will be obvious that the Indian social design provided economic viability to each group. As such, the economic structure was designed on the basic principle of distribution of resources, not distribution of income. Distribution of resources was tied to the configuration of lifestyle, identity, technology and world-view.

Segmentation by technology in the same economic profession also ensured economic roles. For example, there were three kinds of mali (gardeners). Like the weavers they too displayed social distance in terms of intermarrying and interdining. They also displayed differences in their myths of origin. The role tasks were *phal* mali (fruit gardeners), *phul* mali (flower gardeners) and *shak* mali (vegetable gardeners).

The roots of this kind of specialisation go deeper than those of modern professionalism and specialisation, for they were genuine attempts to preserve the total ethnic and psycho-cultural identity, orientation, lifestyles, beliefs and culture within the socio-economic design. This is the basic process of formulation of *jati* or caste.

Self-sufficiency

The Indian social design was congruent with the economic design, based on the principle of distributive authority to each village unit, which was made to rely optimally on local resources. The consumption pattern of various groups avoided dependence on outside resources. Commercial exploitation of distant resources was decided by consensus within the unit. The underlying principle of this governance was creation of equality in membership. There is evidence that even the kings could not impose legislation without consensus. The basic community processes in each autonomous unit or subregion were managed through consensus. The authority of the kings could not be directly exercised in social and economic spheres. Each member unit, however low it might be in the social task role, had to be consulted, influenced, persuaded and negotiated with for change in community action. This was a natural corollary of the ethos which assumed that each individual's life was sacrosanct and could not be violated and interfered with by the power structure and external authority.

Governance by consensus highlighting equality reinforced the principles of distribution of authority. It also triggered off processes which counteracted against the propensities of structures to become hierarchical and allow concentration of authority and power at a single point in the hierarchy. Among these was a lateral system of exercise of authority on social issues. This system came to be known as the panchayat system.

Caste and panchayat system were two sides of the same coin. In tandem they proved effective in society. To hold them together and retain social efficacy, the *jajmani* system evolved. This system was not a mere exchange of services or barter. It was anchored in the construct of distribution of resources. For example, a family could not dismiss their scavenger for inefficiency and neglect. If they did, the village panchayat intervened to censure the family and restore the scavenger to his family heritage and privilege. The head of a certain family, being somewhat modern (in 1953), ignored the censure and refused to take the scavenger back. But no other scavenger came to work. The scavenger community, with the panchayat's support, filed a case against the family. The law courts decided against the family. The judgement was based on the concept that the scavenging service is an economic resource for the scavenger community. The family in question, hence, was a resource to that particular scavenger. Resources, as distributed by social codification, cannot be taken away.

Social Codification

In the Hindu social design there is no evidence of jurisprudence and a legally defined system of punishment. Most of the punishments are recommended. The unit communities examined offences against the social code in the context of events. The *panch* declared a restorative punishment which was implemented through social censorship rather than legal force. Even the determination of the restorative punishment was open to negotiation during the process of achieving consensus.

The second-level boundary conditions were aimed and directed at creating self-contained and self-sustained unit communities. This allowed the members of the community an optimal social and economic existence within the geophysical resources of the locale of the community, leaving them free, if they wished, to work for the ultimate goal of *manan* (spiritual reflection) and self-realisation.

Continuity of Psycho-social Relatedness

Such a social design recognised no discontinuity of the secondary system from the primary system. The voluntary relatedness and the complex pattern of familial roles were repeated in the two secondary systems of caste (occupation and education). The caste distribution (occupational hierarchy) of resources was also involuntary. An individual was born in it, as he was in the family. The nomenclature of familial roles was repeated for the caste elders. Psychological attitudes of relatedness were also reproduced in caste relationships. Group in-fighting in the caste hierarchy was a replica of squabbles in the joint family.

The education system, for those privileged to avail of it, demanded physical separation from the family. The individual left his home and entered his teacher's home to which he carried the same attitude of psychological relatedness as in the family. The guru was the father and his wife was the mother, and peers were *gurubhais* (brothers). Upon starting work, the individual entered the producer–society in which there were occupational associations independent of caste. The individual returned to a system which was qualitatively the same as that in the family. Essentially the socialisation of the individual in secondary systems was on the same pattern as in the familial systems.

This aspect of the secondary system provided the basic cohesive forces of unit-based social organisation on the one hand, and stability of psycho-cultural identity for the individual on the other. Hence the individual did not have to go through a psychological moratorium in the life-cycle. There was no need to reflect, reassess and redefine society in terms of the emergent identity. The wherewithal and mechanisms for relatedness in the family were adequate to last a lifetime. While the process provided stability to the psycho-cultural identity, it tended to overdefine the functional role identity and its linkages with the system.

Socio-psychological Infrastructures of Life Space

The other level of boundary condition in the social design related to the socio-psychological aspects of life. Essentially, each individual, wherever he might be, was in the process of unfolding his karma and going through the process of realising his infiniteness. While his life boundaries and his identity were sacrosanct and inviolate, he could only be given a task role in the social design. (The individual then had to be conceptualised as being infinite and finite simultaneously.) His finite existence could not be treated as his totality. Hence, the social task roles defining his transient finiteness could not bind him completely. The assumption that the individual is simultaneously a person, a self and a social role made social designers realise that such a social organisation would inevitably end in self-role conflicts. Therefore, mechanisms for resolving the stress of self-role conflict, temporarily or permanently, had to be introduced. This resulted in the three boundary conditions defining the socio-psychological infrastructures as freedom in worship, beliefs and rituals.

Freedom in Worship, Beliefs and Rituals

Although social designers strictly codified the social task roles and laid down norms of behaviour and boundaries of interaction of the role, they left the person free to express his being in a specified space and through rituals. These devices allowed for the expression of those instinctive and ontogenetic impulses which would otherwise disturb the social organisation. In this freedom even caste distinctions vanished. These specified locations for rituals became the socio-psychological space for the individual to explore and experiment. In terms of modern psychology the psychopathological space provided outlets for psychopathological tendencies without creating any adverse effects on the society.

Examples are abundant: a male with transvestite inclinations could join the Sakhi Sampradaya, and dress and act as a female within the boundaries of the cult. An intensely exhibitionist male could join the Nangas. A person who needed to emasculate and castrate himself and overcome his anxiety could worship Bahu Charaji. A person with psycho-sexual problems could find resolution and sublimation in the various subsects of Vama Margis and Shakya. There was even scope for abandoning the mask of culture to fulfil needs and physical impulses in these spaces.

Festivals

Festivals, as part of the social infrastructure, were another mode of expressing residual frustrations. Clustered around the main religious ritual of the festival, various communities across the country developed varied activities through conventions. These activities performed two functions—discharging reactive and residual feelings and allowing expression for actions and feelings tabooed in actual role relationships. Holi with its various rituals is perhaps the most potent and clear evidence of the functions of festivals. The main ritual celebrates the courage to act from conviction, even in the face of possible threat of loss of life. Around this main ritual are woven conventions expressing the joy of harvest and the thrill of the advent of spring with playfulness and erotic songs. There are also conventions breaking down taboos on the expression of residual feelings. The example of temporary breakdown of taboos is most clear in the horseplay between the brother-in-law and the sister-in-law. All the year round, the relationship between the two is sanctified; it is a mother–child relationship. She is supposed to act as a buffer, softening the maltreatment meted out to the brother-in-law by the significant roles in the family. But in public through the horseplay of throwing colour on each other, they can freely give vent to their suppressed erotic feelings.

The expression of residual reactive feelings can best be illustrated with the following examples. One custom in Bihar concerns young boys between eight and twelve years of age. They form a group and move from house to house singing loud songs. Each boy strips completely when he reaches his own house. When he knocks at the door, his father opens it, (if there is no son in the family the nephew takes up the role) and listens quietly to the torrent of abuses which the boy utters. At the end of this outpouring the father pays him some money, and the boys go on to the next house. In Punjab, Haryana and the Brij area of Uttar Pradesh women go outside their homes and block the streets. They carry sticks, and any male who dares to pass along the street receives a token beating, sometimes even a heavy thrashing, from them. Incidents are not uncommon where the thrashing has been so violent that wounds are inflicted and blood drawn.

In small towns or villages of North India, people throng the houses of the most important people in the community who, by virtue of wealth and status, enjoy excessive power in the social system, sometimes using it for their own ends. The people ask one such person to come out of the house, and then blacken his face and garland him with a string of old shoes; sometimes even making him ride a donkey while taking him out in a procession and hurling all sorts of indignities at him. The educated, scholarly and intellectual community hold an event titled, 'Maha Murkh Mandal' (The Society of Dunces) wherein individuals are given titles reflecting their idiosyncracies and proclivities; these are even published in newspapers.

Whereas Holi is perhaps best illustrative of this carnivalesque function, a proper look at any festival such as Makar Sankranti provides evidence of how a whole set of conventions around festivals creates space for expression of residual feelings or tabooed impulses.

Freedom of Choice between the Role and Task Self

A religio-philosophic ethos, centred around self-realisation as the main purpose, allowed the individual to surrender the role without feelings of guilt or fear of punishment. Indian social history presents a number of individuals who abandoned their roles, followed their own convictions and designed a lifestyle and space of their own. Buddha, Mahavir, Shankaracharya, Sant Sukhobhai, Chaitanya, Meerabai, Surdas, Tulsidas and Raidas are some of these. Either total or partial abandoning of the role was permissible. For example, Gandhi gave up the conjugal aspect of his role as a husband. In fact, examples of abandonment of a part of the role, whether that of a husband, father or son, are more frequent than one realises.

Part of the Indian societal organisation was thus designed to sustain such an ethos, anchored in institutionalised forms which permitted the individual to follow his unfolding and yet preserve the stability of society. Through the second-level boundary conditions, it created a setting and a social fabric where individuals in limited time and restricted space could work towards their spiritual goals. It thus provided a unique individualism and individuation, within the setting of a well-defined social role. The definition of the social reality is implied in the cardinal principle of duty, i.e., dharma. However, duty was not coercive in nature and not in the nature of a 'must'. It was more of an 'ought' and the individual was always left free. But he always had to face the classical conflict, like Arjuna, 'What is *kartavya* and what is *akartavya?*' (What ought to be done, and what ought not to be done?)

In some historical periods, therefore, the Indian social design provided freedom for some to criticise openly the overall social reality. It provided space to institute different and unique beliefs, values and lifestyles. It provided space for continuous reinterpretation of ethos and behavioural forms. All these go to support individualism as a fundamental value of Indian identity. Few modern interpretations recognise this.

Most other social organisations, primarily those of the West, are designed with the specific purpose of creating a coherent collectivity where the expression of only one set of psychic impulses is permitted. The other part of the duality has to be suppressed, conquered or sublimated. Failure to achieve this results in a sense of guilt, whereas the Indian social design utilised the intrinsic ambivalence of the human psyche. This utilisation created simultaneously coexisting spaces for all the three monologues of existential ambivalences, which are:

I am, I am not—who am I? The dilemma of Being
I can, I cannot—Can I? The dilemma of Choice of Action
I do, I do not—Do I? The dilemma of Involvement

The process created a space for being a role, a space for denying a role, a space for action, a space for role tasks and a space to deny aggression, eroticism, manipulation and impulsiveness. Simultaneously, it designed a space to express these very feelings through ritualistic and symbolic behaviour, and also through direct action under the mask of experiencing one's own infiniteness and one's own salvation. It designed action spaces to explore and experiment with many aspects of the being which were excluded from the role task space. And finally, it delinked the action space from the intention space by postulating detachment on the one hand, and the status of being a medium to the self on the other. Thus, the individual had the choice to treat himself as an agent of action or as a medium of action.

The Indian social design created simultaneously the forces for socialisation and the forces of individuation; the former pushed the individual to conform to the absolutism of role behaviour, demanding a surrender of the self and unquestioned commitment to withhold personal feelings from action. Rama and all the other characters of the *Ramayana* epitomise conformist behaviour. The latter, the forces of individuation, left one on a razor's edge. They required making constant proactive choices while giving attention to situational variances.

These are crystallised in the role of Krishna and his teachings. It is interesting to note that in the *Mahabharata*, Bhishma, Drona and others who separate personal feelings from action, in order to meet a rigid role commitment, stand opposed to Krishna who represents the ideal of individuation.

The social design became the basis for constructing the socio-psychological world of objects and a symbolic world of meanings and concepts. It led to a mode of pattern–thought rather than linear thought. Every sphere of knowledge was interconnected with every other sphere of knowledge of life. Every object was comprehensive and multifunctional. Most words were not mere object references, but connected a whole universe of meanings, behaviour elements and choices. Many of them in the Indian tradition were contradictory. For example caste, through stratification and distribution of power, represented the dilemma of quality of personhood and hierarchy of status in its political aspect. Socio-economically speaking it demarcated an occupational space, achieving economic segmentation. This was based on technology on the one hand, and ethnic myths on the other. It bound the field of interpersonal relations with the consequential dilemmas of inclusion and exclusion in relationships. In the socio-psychological sense, it represented the differentiation between self and role tasks. The concept represented simultaneously many levels and aspects of social living. Individuals, in taking a stance with regard to the concept, were always on a 'seesaw'.

In the object world, a similar phenomenon of patterned thought can be illustrated by the most commonplace object, the *khat* (bed). The *khat* is not only a bed but it is also a sofa for receiving guests; a table-cum-chair for eating or playing cards; a rack to store things upon; a space to dry clothes and process food; a prop and a ladder to climb walls; a shade against the sun; a shelter against wind and rain—in fact it is the most versatile object in its use.

Similar examples abound in the field of knowledge. A *jyotishi* (astrologer) knows something about medicine, weather, food, gems and other things related to the wellbeing and future of an individual. A musician is not only an expert in vocal or instrumental rendering. His training involves understanding the intricate theory of emotions and the analogous universe of the *swara* (notes) with other aesthetic experiences such as colour, taste and smell. Similarly, a medical man is not only a diagnostician and a prescriber of drugs but has integrated knowledge of *jyotishi* as well as of food and its qualities related to various states of sickness. Essentially, there was no specialisation in knowledge. It was more often holistic. Specificity of expertise could be valid and viable only if there were an awareness of the contextual universe of man's social, psychological and moral existence.

A similar phenomenon of pattern–existence applied to words. Here each word connoted not only the object in its physical existence but also the many levels of the universe of experience with it. For example, there are seventy words to connote the moon as object. These words are neither synonymous nor interchangeable. They connote the quality of experience the moon can evoke. For example, the word *rakesh* represents a whole universe of a full moon seen in a clear blue sky spreading soft, mystic, somewhat chilled light, on a calm night with no turbulent winds. As against *rakesh*, *rajneesh* connotes any moonlit night from the first to the fifteenth of a month. Such words were pregnant, carrying a hierarchy of meanings, some evolutionary and some operational. For example, the word *purush* connoting man, has three meanings. Each is different in genesis and application.

An Overview

The Indian social design was cohesive and comprehensive. It encompassed a wide, kaleidoscopic universe which allowed the elements of social life to form new patterns from era to era. However, it must be emphasised that the ultimate burden of the maintenance and sustenance of the design depended on the individual who, though bound to a limited and restricted space, enjoyed much discretion and choice. This process of locating the final burden on the individual meant his having to walk on a razor's edge. He had to choose between playing the role, being the self, or integrating the two. The role of the intellectuals was to provide periodic reviews of the social processes and reformulate new boundary conditions. It was the most critical variable.

This religio-philosophic ethos and social design once supported a dynamic society throbbing with life and culture which progressed in almost all spheres of life except, perhaps, in the field of technology. This society showed innovations in the field of knowledge; it perfected cultural forms, dance, music and drama, and excelled in the production of consumer goods; initiated and fostered meaningful and profound religious and social thought. In its prime, it absorbed many assaults and impingements from alien sources and yet maintained its character.

The last two centuries have seen a decline of this ethos. The social design has become battered and ineffectual. It has almost become an inverted image of its former existence, and can be best described as a stagnant society, slowly degenerating. Compulsive conformity, role-boundedness, denial of personhood, abrogation of representativeness in the system, dependence, deprivation, discrimination and exploitation have spread throughout the fabric of society, and have made inroads into its culture. Interference, manipulation and aggression, which violate human dignity and crush the human spirit, are rampant.

The society and culture are no more multifaceted. Openness, creativity and innovativeness are slowly disappearing from society and its culture. India has become a borrower society and has lost its acumen, its pride, self-sufficiency and resources. Its past strength has become a dream and its problems have become unmanageable. Indian society today has truly become a pathological version of its bygone days.

In the ultimate analysis today's reality is what matters. Our attempt to relive the past was only a journey comparable to that of Odysseus walking through the portals of Hades, in order to come to terms with the ghosts of his past; to return to the reality of the day as a chastened and reflective human being. We believe in dedicated action and this journey was a way to discover the strengths from our tradition to cope with the current scenario. The journey helped us to deal with our self-hate, shame and our guilt for being born Indians and belonging to India. We found ourselves willing and ready to face the reality of today with understanding and courage. The journey regenerated us as it did the heroes of universal myths. Yet the question remains. How could such a transformation have taken place? What were the elements that converged to create it? What kind of society existed in India so that the encounter with an alien ethos could have such a serious impact?

Three

Reliving Indian Reality: Cognitive Maps from the European Ethos

An Indian today is a child of two cultures. He struggles with two contradictory but synchronic worlds he lives with. One as described in the preceding chapter is the emotive map deeply anchored in the culture, tradition and social system. The emotive map requires the individual to do his duty as prescribed by society, considered socially desirable, and according to the demands of the older generation. The other is the cognitive map of logical thought, rationality and aspirations which creates a world-view quite in tune with the times but out of harmony with the emotive map. This forces the individual to either conform or rebel or just walk away. The Indian identity thus often remains elusive.

Our attempts to discover what this identity is and what are its antecedents led to the exploration of the individuals' cognitive map of socio-psychological infrastructures and their action choices. The explorations of these infrastructures form the base for reconstructing the cognitive map of Indian society as it has been drawn during the last seventy years. This chapter documents the cognitive maps and their bases as held by the participants. It also documents their attempt to locate themselves in the current flux of Indian society.

The explorations of identity and the accompanying socio-psychological infrastructures of choice and action ranged over many aspects of their lives, from lack of communication with parents, siblings and peers, to lack of communication in marriage and eventually with people in their organisations. The explorations ranged from conflict between personal needs and the needs of the family to conflicts involved in choosing a mate and a career. The problems of mutuality, reciprocation, inclusion and exclusion, exploitation and deprivation in relationships continued to haunt the individuals. Issues of commitment and idealism were pitted against those of survival, both social and economic. Problems of guilt, shame, anger and pain contended with creating space, autonomy and meaning for the self. Individuals expressed concern about purpose and direction in life. Fears and anxieties about living mechanically and without zest were voiced. We found individuals concerned with success but holding fear of failure within themselves. Deeper explorations into the dynamics of choices and actions revealed their sense of inadequacy, fear of failure and invalidation, the sense of dissatisfaction with the self and the system, and finally and surprisingly, the sense of poor self-worth.

Out of the explorations of identity emerged a son, a victim, a martyr, a wanderer, an exile, an orphan and a spectator. These patterns created their choices of role taking. They varied from Hanuman, who had immense strength and resources but could use them neither for himself nor for the system; to those always awaiting orders from someone on how to use them, like Parshuram, who in a rage destroyed a whole caste many times over and in reflection retreated into passivity; and Vishwamitra who could be great only in his reaction and in negation. In Greek myths we found Hercules who continued to perform the hardest tasks without receiving his heritage; Atlas, who carried the burden of the world, Sisyphus, who went on pushing boulders while never reaching the top; and Tantalus, who created wonderful things for others but never received anything in return.

Women in their role taking reflected the symbolic identities of Sita who for her devotion to the role was rewarded by loss of home, periods of exile with misery, and separation; mistrusted, she carried the burden without support. Other role models were Padmini whose physical beauty made her a sex object and who found herself at the centre of feuds fought over her person; Parvati, whose husband was either self-absorbed or sensual, but displayed no interest in her personal and social preoccupation; Kannagi who had to fight not only for her own survival but also against the injustice done to her husband; and Meerabai who withdrew into asceticism because no social relationship was acceptable to her. Besides these identities based on mythological and cultural narratives, both men and women displayed identities based on folklore.[1]

The current generation presented its social environment as hostile, exercising excessive control, and devoid of affection and love. They

[1] Indira J. Parikh and Pulin K. Garg, *Indian Women: An Inner Dialogue*, New Delhi: Sage Publications, 1989.

painted a picture of an environment allowing them no space. Space could be created only through conformity or manipulation. They spoke of an environment where expression of any personal feelings or action for fulfilling inner needs elicited anger, resentment and punishment. When challenged to act according to their convictions, they were hesitant to take the risk. Their perception of the environment and the social system was tinged with anxiety and fear. They saw themselves as powerless both to act upon the environment and to replenish it. The opposition of self versus environment was deeply felt. They saw themselves as having no resources to initiate new responses. They had preferred solutions and prescribed choices to meet their problems. Their unwillingness to create solutions was frightening. Passively they withheld their resources from themselves and the system, engaging in extracting advantages for themselves in dubious ways.

Our discussions raised many questions: What is happening to the current generation and why? Why is the process of becoming so tortuous and annihilating? Our articulation of the Indian ethos and social design, as presented in the earlier chapter, made the first dent in the deep-seated scepticism regarding our Indian heritage. Discussions brought to the surface their resentment of the current overemphasis on role behaviour which was the major impediment thwarting self-realisation. It was apparent that barriers still remained around the social system and its infrastructure, which hindered the discharge of residual feelings and taboos. In the light of their experience, the Indian ethos and its accompanying social design are no more what they are claimed to be. The process of transformation has vitiated them and reduced them to the pages of history.

Indian ethos and social design by their very assumptions fostered a synthesis between agrarian craft and the network of services provided within the framework of the caste system. This synthesis integrated both sentiment and task interdependence. Over a period of time the growing distance between ethos and social design, due to the failure of cultural reformulation, converted Indian society into a typical agrarian and role-bound one.

The Ethos of Renaissance

The participants attributed the mindset of modern India to the introduction of English education and technology in the 19th century by British administrators. Indian intellectuals responded to post-Reformation empirical thought and Christianity in diverse ways.

The new ethos based on this was more easily accepted by those in India who had received the new education. The new ethos was assimilated and internalised through its revalidation from the authority of indigenous religious texts. Then the tradition began of validating Indian thought, ideas and expressions by establishing either direct or analogous evidence from the West.

For a time the process of assimilating Western thought brought about a spurt in intellectual and social activities. A glorious Indian past was constructed in the hope of combating the growing inertia of the role ethos anchored deeply in the agrarian mode. However, the hope was premature. Intellectual and social activity failed to stimulate any significant change in the processes of society. Only the content and form had changed and were reorganised.

The generation which entered its youth between 1880 and 1920 seemed to have woven most meaningfully the strands of European thought into the fabric of the Indian value system. This was evident in the fields of politics, literature, physical science or engineering. Both Western ideals and socio-cultural processes were successfully applied to Indian situations.

In order to understand the dynamics emerging from the weaving of the Indian social ethos with that of the West, it is important to identify the elements involved. Enlightenment, emancipation and progress became the keywords for action. The woman's role, the Hindu joint family, the caste system, and in fact all the institutions of differentiation and integration within the Indian social design became targets of attack. Social justice and democracy were other important values of the Renaissance, along with the spirit of scientific enquiry, to which the Indian elite responded. This gave birth to new cognitive maps of people and society. Rationality, logic and consistency became the anchor words.

Ideas regarding equality, democracy, social justice, education and scientific enquiry became the stimuli for reconstruction of the social fabric of Indian society. Rationality, objectivity, concern for the fate of the common man, and universalisation of education became commitments for action.

It is interesting to note that this ethos did not influence the common man unless it had the backing of Indian thought. The success of Dayanand, Gandhi, Raja Ram Mohan Roy and Tilak testify to this.

Dayanand, the founder of the Arya Samaj, created a ritual-free Vedantic philosophy. To promote the concept of swaraj, Gandhi invoked the image of 'Ram-Rajya'. The new ideas led to an intellectual ferment among the educated upper classes of the society. The new ideas became the goals of social action for which associations and institutions were set up. However, their actual functioning was in continuity with the Indian agrarian ethos. Thus began another tradition of adopting forms and borrowing contents from the West, without designing and institutionalising infrastructural and sometimes structural processes.

One of the participants said:

> It led to new challenges and disowning of the old commitment in each of us, but our old expectations of the system continued. We imposed new demands but never recognised the new commitments that were needed for a real graft. They remained unknown.

This comment gives a glimpse into what actually happened when the new ideals were promoted in practice. The participants quoted many examples of anachronistic forms and processes. One example reflected the exercise of authority and decision-making in familial as well as organisational settings. Almost no decision could be made at the task level by the role holder. All situations were converted into problems. The process was to surrender the actual decision-making to one person. The motto was 'passing the buck upward'.

The interdependence between tasks and the Indian social ethos, interwined in the caste and *jajmani* systems, had been eroded. The erosion gave rise to the 'exclusion–inclusion' processes among the members of the collectivity. Eventually the relationship between classes of members became governed by a kind of absolutism of prescribed role boundaries. The concepts of equality and fraternity were not much in evidence. They could only be sustained through legal codes of conduct. Indian society discovered many ways through which the traditional casteist processes could be continued. The beliefs and values of a traditional ethos held sway in private lives. In public life idealism of the logical rational mode was articulated. Individuals were caught between the two modalities. Personal choices based on values and convictions became difficult. Only a few could rise beyond the clash of the two modalities to act with conviction.

Individuation promoted by the Indian ethos of the past led some individuals, including Gandhi, to act with conviction. Action was inevitable in this commitment. The adoption of a different ethos did not release or create a new infrastructure for action, in spite of the fact that in its original form it had promoted new actions. In India 'equality' and 'fraternity' became slogans for awakening and creating social awareness and criticising prevalent social practices. However, the responsibility for taking initiative and action was left to the external authority. The concepts of equality and fraternity became instruments for mobilising social authority through pressure. Individuals did not act by choice or according to those values as was the case with Indian ethos.

Thus, the ethos emphasising equality, fraternity, liberty and social justice fused with the cognitive orientations of the Indian ethos. However, fusion with the emotive or action orientations did not occur. There was a lag between them. Indian culture had failed to generate the required network of attitudes, values, and beliefs. This mismatch between cognitive orientation and emotive and action orientations became one of the critical loci of disintegration. Each orientation became a compulsion, pulling individuals in different directions. The question was how to integrate these three orientations. Most individuals, instead of attempting to resolve the conflict, blamed the significant roles of the system for creating it. It then became easy to take the next step of questioning the integrity of the system. Buffeted by unresolved and seemingly insoluble dilemmas individuals, sooner or later, turned their backs on the system and went their own ways.

The imported cognitive orientations, however, were used by the elite to awaken the masses and organise a struggle against the political enslavement of India. They also generated a few reform movements which helped in the construction of India as a nation, as against a psycho-cultural system. This was nothing new. Even in Europe, the Enlightenment—being an embodiment of values which would free a society caught up in the morbidity of tradition and role-boundedness—was a cognitive instrument for breaking up the feudal system. It was only when the technological, i.e., the applied, aspect of knowledge grew dominant in the West that such ideals for action gained currency.

By the mid-thirties, the nationalist liberal movement was losing its impetus in India. It was espoused by a few individuals like Nehru. It was partially replaced by the Gandhian ideology for the common man, which was akin to the religio-philosophic ethos and social design of India. The Gandhian movement differed from the nationalist liberal one by laying down a bold and innovative social design in the form of constructive work. It retained the village as a unit, with the accompanying principles of distributive authority and governance by consensus and the mechanism of influencing rather than exercising authority. Gandhian ideology also retained the model of a producer society using small-scale technology. It attempted to break down the boundaries of caste and profession by choice rather than by birth. In many other ways, Gandhian ideology was a comprehensive and renovated model of primary society in India. In the field of social reconstruction the movement was in the tradition of the past. Intellectuals worked through discussion and consensus. They identified and diagnosed the dysfunctionalities of society, did a prognostic analysis, and initiated remedial action. It was surprising for us to discover that this significant aspect of Gandhi's approach to building a new India was either unknown to or ignored by the current generation. Gandhi came to be seen as a political instrument of Nehru's dreams of building a modern, technological India.

Another European ideological movement which made a considerable impact was Marxism and its social design. Many young men and women took the trouble of understanding Marxism to find out whether it was an appropriate design for Indian society. It is a tragic commentary on Indian education that the youth studied in depth the Marxian ethos, together with the communist social design, but were never exposed to a systematic understanding of the Indian and Gandhian ethos and social design. It is one more example of the fall-out of English education on Indian institutions (Garg and Parikh 1976).

The Second World War was followed by the political independence of India which saw the Gandhian ethos sidelined. Nehru—the elite charismatic leader, young at heart, the modern Indian Ulysses—was thoroughly steeped in Western liberal and scientific thought, in spite of his reflective journey in *The Discovery of India*. He chose European society as a model for India's technological development. There was an almost exclusive focus on goals of and means for material development. There was no investment in or even an attempt to build a new morality, goals and processes. Nehru focussed on action. When called upon for a new socio-political philosophy he repeated the Gandhian ideals and the ethos of reform. Nehru referred to the dams and hydroelectric power stations 'the temples of new India'. This shift in emphasis brought the technological aspect of the West to the forefront, which so far had only been a slow-moving companion of its ideational ethos. The urgency of development made Nehru and the new leadership ignore the need to rebuild the ethos and create new infrastructural processes. This neglect—a failure in our opinion—has been damaging to the psyche and identity of the current generation. This is what makes them sceptical and desperate on the one hand, and anxious, tense and frightened on the other. They have no grounds on which to base their faith and convictions, only reasons and justifications to back their choices.

The Scientific and Technological Ethos

A large number of people representing our sample (taken during 1972–86) were born between 1948 and 1966. A small section was born around 1940. And the third set of participants (studied during 1976–86) was born in the late 1950s and early 1960s. Most of them grew up during the period of decline of Gandhism and liberal humanist reform and the dawn of the technological era. They absorbed the developmental ideology of Nehru and like him put their trust in science and technology.

Our dialogue with them brought home to them and ourselves the mortifying discovery that the current generation had been denied an exposure to the religio-philosophic ethos, an awareness of the social design, and an understanding of the heritage of the past. They were ignorant of the immense social and existential suffering experienced during years of struggle against the colonial yoke. They were not sensitive to all the 'native' pathos so beautifully and wonderfully portrayed in the literature, prose and poetry of the time in various Indian languages. They were blissfully unaware of Indian heroes, myths and folk-tales. Comics, Enid Blytons and the popular Westerns, adventure tales, detective stories and spy thrillers formed their staple reading. They had not even been exposed to the classics of the West.

With this technological thrust another window to the West was opened, in particular to the USA which soon replaced Europe as a reference model. This changed perspectives, and rather suddenly. The techno-economic, eco-political and scientific coordinates of another reality, a vast melting-pot, were imported in totality to guide national development. Input–output models became dominant. Individuals and society were then treated as a black box. When the input–output model, determined and implemented by the abovementioned three coordinates of ideality, failed or produced only partial results, the black box of culture and society was blamed. Thus another period of dissatisfaction with one's life and nation unfolded. The current generation looked beyond them to the lifestyles portrayed by the media in the West.

The parents and the educational institutions emphasised acquisition of techno-informative knowledge and achievement of high grades. High concentration on role performance and optimising career or occupational opportunities preoccupied the current generation. This process excluded building a broad perspective and engaging with the realities of the system. They grew tall in competence but were very poorly grounded in socio-cultural processes. They limited their action choices within occupational frames. They solved day-to-day problems and suffered inwardly the stress of a routine social existence. The overemphasis on academic role performance precluded enjoyment of many other dimensions of the life space. In effect, a childhood was denied. The first group remained being good sons and daughters; a later generation born around 1940 got hooked on to fads, creeds and consumerism on the Western pattern. At the same time, they became resentful of the dynamics of the social system and their personal lives. The new horizons of developmental activities provided hope for some, while others experienced nothing but despair. As one of the participants remarked, 'Things loom large on the horizon. They appear to be coming close, but eventually fade away.'

Scepticism became the hallmark of this generation of the 1960s and 1970s. Perhaps scepticism masked the impatience, hurry, fear of being left behind, unrealistic equating of oneself with the elite, and a belief in their capabilities to earn every reward society could offer. At the level of the self, the suspicion of one's limited abilities led to feelings of deprivation, discrimination and unresolved sibling rivalries. In educational institutions and formal work organisations this led to competition, conflict and comparison.

These feelings made sceptics out of the first wave of the current generation; the decision-makers and planners continued in their attempts to emulate the borrowed models of the West and blamed the culture and the people for the limited success of these models in Indian society. This first wave chose the road to the right, the road to conformity with the existing social context. But the rest of the country remained psychologically incapable of reacting to the new directions introduced by the visualisers of new India. It was left to the second wave, the new generation, to choose the second road to conformity, again borrowed from the West. They picked up forms, fads and creeds and indulged in self-centred living. The first wave, though sceptical, struggled to seek values and meaning. The second wave partly disengaged itself from the system and sought catharsis in being 'modern'. Some others turned to radical ideologies and planted the seeds of disruption and violence as means of managing non-belonging. They cut themselves off from hollow commitment to the system. They learnt to shout hollow slogans which gave some direction, meaning and goals to their empty lives.

The second and third waves of the current generation found their life space dominated by the new techno-economic, eco-political and techno-scientific forces. The introduction of modern technology into India then became only a means for increasing production, introducing new products, creating resources, and raising the standard of living of the masses. Like all major manifest changes in human systems, this choice also had unintended consequences for the Indian social design, its ethos and the philosophical assumptions of man and collectivity. When this impact started becoming cumulative and apparent, Indian intellectuals suffered, being influenced by Western concepts of alienation, dehumanisation and isolation at the individual level. At the group level they became aware of disruption of harmony, cohesion and patterns of social relationships. In the cultural context they found disintegration of values and institutions and a rise in unhealthy conflicts. From the economic point of view, the intellectuals accepted the resulting imbalance in the distribution of wealth, economic buoyancy, and rise in the consumer-oriented economy.

These Western models became the standard explanations for the emergent social phenomenon in India. These explanations made sense logically but did not lead to any cultural change. The urgency to catch up with the West added another dimension. Within the lifetime of the current generation Indian policy-makers pushed ahead with a shift from mechanical technology to process technology and finally to high-tech. A process which took several decades in the West was compressed within two decades in India. To a certain extent the accompanying ethos of the mechanical technology had evolved in India. The attempts to introduce and compress the newest technologies within a very short time created disruption in the context of living, social relationships, modes of meeting life situations, as well as the existing social and work infrastructures of society.

If we look back at the history of technology, it suggests that the technology of production is one of the significant parameters of the life space. It needs a congruent ethos and social design to be effective. This important link between technology of production and technology of living has often not been understood. Technology has been conceived as an instrument in the hands of man. The Indian social designers believed that inherent in any technology are images of man, the collectivity and the relationship between them. Significantly, the technology of living requires different kinds of psychological infrastructures of action and relatedness. These have to be visualised, designed, planned, built and made operative. Delay in the formulation of these infrastructures creates a culture and process lag, which can only increase the spread and intensity of the unintended consequences and eventually lead to dysfunctionalities and pathology.

The failure to build congruence between the social ethos and technology has generated forces in almost all Third World countries of fragmentation and psychic uprooting, anchoring life-styles in consumer-oriented fundamentalist movements and/or military regimes. Indian society has reacted with increasing consumer-oriented behaviour and has also continued to be fragmented by fundamentalism, religious movements and/or parochalism and sectarianism. All these have tended to converge in the socio-religious processes of India and have acquired a strong hold over the people's minds.

The manifestation of these is also apparent in the social context. For example, revival of dowry, massive resurgence of rituals, founding of new temples, convening of religious conferences, reemergence of intercaste rigidities and conflicts are some of the glaring instances. All these reflect the failure of Indian social planners to design appropriate psychological infrastructures and social institutions to counterbalance the unintended consequences of technological development.

The intended or the unintended choice to follow the path of the West as projected by the media led to several consequences which present themselves in unique patterns. Let us look at some of the process-level consequences.

Mobility

Mobility, physical and social, was the first major consequence of the technological development. It cut into the fabric of Indian community life. It brought into sharper focus the rural–urban, agrian–industrial, the haves–have nots, and the literate–illiterate divides. With the passage of time mobility generated patterns of social and personal conduct which ran counter to the traditional Indian social code of conduct and behaviour. Under the impact of the new technology, the existing socio-cultural processes and strategies failed to unfold or cope with the consequences of mobility. Each individual was left to his own devices to handle the situation.

The technological thrust led to a process of migration of rural people to urban centres. In the initial stages, the migrants from rural communities to the centres of the technological systems perceived themselves as extensions of the family left behind in the rural setting. For them the movement was only a physical one. The migrants took over the role of income and resource generators for the family. They kept alive the myth of returning home some day. Thus, they established a house in exile, and not a home. The second wave of mobility created migration patterns across state borders and eventually to Western shores. Many Indians continue this process even today as they migrate abroad. This process too had several phases. In the initial phases the migrants who went to the West also held on to the belief of returning home. As time went on and they put down roots, it became evident that returning home was a dream. This process denied psychological mobility. Psychological mobility and self-imposed exile remained associated with guilt, shame and remorse. The migrants to the urban areas and abroad failed to integrate themselves with the local community. They rarely, if ever, put down psychological roots. They hardly ever invested and replenished the cultural and social life space of the community in which they found themselves.

In the last decade a similar process has occurred with the Gulf migrants. Unlike the first phase, some of the dreams of building a home back home has come true for many. However, while in exile they have remained as aliens.

The socio-psychological symbiotic fabric of the Indian community and society began to fray at the edges. Increasingly self-oriented living, which placed primary emphasis one one's own material needs, appeared on the scene. Some grandparents and largely the parents of the current generation in the groups were the first ones to live away from home to carry on their profession. However, emotionally they continued to be strongly tied to their parochial moorings.

Disintegration of the symbiotic social patterns began when individuals neglected their primary families and utilised their earnings for their own immediate family. Nevertheless, they continued to make demands on the family heritage and resources and thought that they had a natural right to it. Mobility, as such, created a split in the mutuality of expectations. The individual began to consider the familial expectations due from him as a burden, but he continued to have expectations from the familial system and felt upset if they were not fulfilled.

The effects of mobility on the current generation are deep and widespread. They dreamt of independence from familial control and wanted freedom through economic viability. Their concept of freedom was freedom without responsibility and independence without commitment. They wanted controls from neither people or systems nor did they want to contribute or be accountable to them. Yet, their emotional dependence remained unexamined and unresolved. Their mode of relatedness to the system was characterised by reactive independence and a sense of being exploited when the primary or secondary systems made demands on them. Distantiation, under the impact of mobility, led to a change in quality of relatedness. Eventually, this got reflected in their fear of closeness and intimacy. Emotional relationships brought out anxieties and apprehensions of being controlled and possessed. Basic trust in normal role relationships was shaken. Their emphasis in most relationships shifted to taking, receiving and extracting. Eventually behaviour became focussed as excessive social activity leaving the individual isolated. Giving of themselves in relationships became difficult, while providing justifiable reasons for making or breaking relationships became a mode of life. Mobility generated patterns of resource related relationships and making use of relationships for personal needs.

Thus, mobility resulted in the individual carving out his economic and social boundaries away from the family. Yet, the process remained incomplete because he only disowned the demands of the system but continued to make demands on the family system. This process reflects only manifest mobility because the current generation carried with them deeply buried models of rural and semi-rural society. They carried emotive maps of exploitation and their accompanying anxieties. As such, only social mobility took root. Psychological mobility did not emerge. Individuals could not initiate effective voluntary relationships. Most difficult was building healthy peer and man-woman relationships. Consequently, most socio-psychological relationships became a space for the intense projection of ambivalence on the one hand, and search for action and evidence for security on the other. Most relationships then, became double binds. Holding on to them was painful, but letting them go was frightening. Eventually, the third wave of the current generation experienced processes of psychological isolation and loneliness. The individual acquired a cactus-like identity.

Breakdown of the Structure of Authority

Technological developments fragmented the space of work and social living and thus disturbed the structure of authority of Indian agrarian society. In the work space the existing and operative authority of social relationships and conduct became eroded by the technological system. A new phenomenon emerged. An individual in the formal work system could exercise authority over those who were his superiors in caste and age. The explicit social authority and status, traditionally assigned by age or caste rank, became formally inoperative. Individuals coped with this disruption of the traditional authority pattern by bifurcating the organisational and social situations. In the organisational setting they accommodated themselves to the operational authority. They suppressed resentment but allowed its expression in passive rebellion such as delaying action, not following instructions fully, and obeying the letter rather than the spirit of the order. The traditional authority structure continued to operate in social situations.

The technological system created a new level of social status, i.e., the peers of lateral collegiate roles. Such a structure of collegiate roles had not existed in Indian agrarian society, where community and task interdependence were harmonised through congruence of the caste and *jajmani* systems. In formal work organisations managing relationships across peer roles needed a reorientation of relatedness and a new mode of exercising authority different from the past. No models existed in Indian society for the newly needed behaviour. Even after one hundred years of mechanical technology and three decades of continuous process technology, relatedness and dealing with authority and systems are critical issues in most organisations. Models of behaviour and skills borrowed from the West have not succeeded, as the supportive network of attitudes does not exist. The building of a network similar to Western models has been difficult as there is no cultural base for it. In traditional, agrarian Indian society technology was either small-scale or simple. Technical competence could be acquired easily by a large number of people. In such a set-up no concept of technical authority ever existed. The introduction of technology from the West also introduced a new concept of technical authority which further eroded the existing mode of social authority.

In Indian society the major concerns were social tasks. Only social authority was legitimised. With the introduction of technological systems, the task of legitimising technical authority confronted Indian organisations. The newly emerging formal work organisations required the simultaneous functioning of technical and administrative (social) authority. In operationalising, however, the administrative (social) authority tended to ignore, overrule or bypass the technical authority because it had not existed earlier. In most Indian organisations the issue of technocrats versus generalists still prevails. The lack of resolution of this issue is classically displayed in the civil services. The IAS (Indian Administrative Service) officer, a generalist, is once again gaining ascendancy. A bureaucrat occupies the role of the executive in some of the most complex technological systems of production and services. He has also started to take over important roles in the education system. He governs himself and the systems exclusively through his perceptions of administrative reality and not task reality. He does not have the appropriate attitudes and organisational processes required for managing complex technological systems. In our assessment an exclusive focus on the administrative reality without enmeshing the technology in congruent attitudes leads to large-scale, invisible waste of national and organisational resources.

The technocrats who are convinced that technical authority is necessary for success seem to have adopted the same modality as the administrative heads. They want to be chief executives of technical systems and take over the administrative (social) authority roles as well. Once they become chief executives they claim authority in all other technical fields, thus eroding the authority of other technocrats.

With the passage of time and increasing focus on technological growth and complexity, the ascendancy of technical authority and professional leadership in task situations began to be consolidated. On the other hand the same technocrats began to claim authority and leadership in social situations. To illustrate, natural scientists who have technical authority in task situations acquire significance through roles of chief executives, directors, etc. They then assign to themselves a natural right to leadership and authority in social situations. The process becomes a one-person system.

Society as well as formal work organisations failed to recognise that both technical and social authority have their place, and need to work in tandem for the functioning of society as whole. This failure created issues of role clarity and boundaries both in task and social systems. To avoid these problems some role holders began shirking their roles, becoming defensive in their performance, or learning to ignore both kinds of authority. Sometimes they created conflicts between the two kinds of authorities to the detriment of the system. Most of them, specially in management systems, became subject to dual controls. This created inertia in decision-making and implementation of policies. The role receiver was trapped with double bind messages. Much of the energy of the organisation was spent in dealing with the problems arising out of dual control.

Creation of New Goals Sets

The third contribution, creation of new goal sets, disturbed the fabric of Indian agrarian social patterns. In the agrarian society successive generations followed the occupational as well as the goal sets of the family or the caste at large. Even when other occupations were more rewarding, people followed the goal sets of the family or caste. This made for stability in social relatedness among castes and families. The infrastructure of the *jajmani* system reinforced the family or caste goal sets. Few individuals made clear breaks. And when they did, it was only by migrating from their homes and communities. As migrant individuals they could accept other occupations and not incur the disapproval of either family or community.

Following the thrust of technology, new professions, and consequently new goal sets, emerged. However, for a long time the cognitive and emotive orientation of each generation remained agrarian. The first generation parents or grandparents at the crossroads of transition opted for new goal sets. As in the agrarian society, they too imposed their new choices on their own children. For example, doctors wanted their sons to become doctors; a bureaucrat wanted his son to enter the civil service.

Initially, this became the pattern across all new professions. Subsequently, the pattern changed. Now each generation was destined to follow one of the goals of the new goal sets. A doctor having three sons wanted one to become a doctor, while the others could become scientists, engineers or lawyers. This pattern continues even today with newer professions emerging on the India scene.

As the monetary significance of these emerging goal sets began to be realised, a channelisation, creating intense competition, towards the same goals began. The growing individuals were left with no choice to scan the environment, weigh the alternatives and choose goals for themselves. They herded themselves towards the overchannelised goals. Monetary incentives persuaded many individuals to step into work roles which had been traditionally taboo for them. However, this stepping out of the traditional work role boundaries made no significant difference to their attitudes to social role taking. Thus the new goal sets created by the technological thrust reinforced personal and social mobility. It also gave an impetus to the creation and development of professional and technical leadership. However, the attitudes of role taking did not change and, therefore, no qualitative change in the relatedness of the individual with his society occurred. Psychological mobility did not emerge.

The current generation struggles with these conflicts created by the process lag between the new technology, professions, goal sets and the existing cultural processes of the agrarian society. Soon it became apparent that the choice of the new professions and goal sets was forced one. The selection of the overchannelised new goal sets was not really in tune with their own aspirations, talents and/or personal assessment of themselves. This created problems of motivation, involvement and commitment. While the new goal sets were embraced by society and culture, the congruent network of attitudes of role taking was rejected. The new goal sets implied some autonomy and participation in system, decisions on the part the individuals, which were not acceptable either to the parents or the authorities in the organisations.

In the present generation we see a struggle for developing a congruence between the new goal set, role taking processes, and the accompanying network of attitudes. The struggle to achieve this congruence is still unresolved. Indian organisations continue to adopt new forms but do not accept the essential processes of new forms. This results in communication gaps between the current generation and the authorities on issues of expectations, values, the nature of the relationship between roles and systems, and the quality of belonging to and membership in the system.

Emergence of Voluntary Relations

The fourth direct contribution of the technological thrust introduced another possibility for conflict in the agrarian base of India. The primary system, consisting of the family and its extensions, the caste and the village community, could no more function as an all-inclusive system of belonging. Contact with individuals with whom there was no primary kinship created opportunities for developing voluntary relationships in educational and work organisations. With time they matured and took the form of kinship relationships. The new goal sets created secondary sources of identification, i.e., with the profession. As a consequence, associations fostering professional brotherhood emerged. These two types of kinships now jointly competed with the primary kinship for emotional investment. Examples are plenty where the natural uncle is given less importance as compared to the 'adopted' uncle from the voluntary, social or professional kinship. For the first time in India, a new system of belonging started to gain significance.

This indirect impact of the technological thrust was once again resolved by Indian society in its own characteristic way. The old cognitive and emotive orientations were transferred to the new system of belonging. The first sign of transfer of emotive and cognitive orientation was in the use of familial terms of address such as 'uncle' and 'aunty' for people in the voluntary system of belonging. These new relationships reflected the interplay of expectations and the pattern of role taking similar to those in the primary kinship. Thus, a voluntary social kinship developed during education and later at work. The process of secondary voluntary systems remaining distinct from primary systems did not develop. As in other aspects of change, here also the forms and content changed, while the old processes continued. A professional kinship developed more out of cognitive rather than emotional need.

The growth of second-level systems of professional and voluntary social kinships generated yet another culture and process lag. The traditional primary systems of India fostered involuntary relationships. These relationships could be maintained simply by holding on to a network of attitudes and processes of role taking. On the other hand, inherent in the voluntary relationships were negotiative processes where expectations could be bound and limited to the reality of the relationship. Understanding boundaries of transactions was necessary. In essence, a process of redefining and redesigning relationships was an inherent dimension of these relationships. The parents of the current generation successfully coped with the new demands in relationships by converting the voluntary social kinship into an extension of the primary family and the professional kinship into a power game of sibling cliques. The current generation is still discovering the processes of role making on the one hand, and the difference in the quality of relatedness between the primary and the secondary systems on the other. Relational orientations viable for the primary system when carried over to the secondary systems make relationships non-negotiable and somewhat oppressive. The difficulty of resolving this issue left them feeling incompetent in managing interpersonal relationships. The fear of not finding mutuality, trust and acceptance continued to haunt them.

Handling diverse, voluntary and involuntary relationships, making a role, and operating from different attitudes and relatedness to the two systems is the greatest dilemma for these individuals. Examples abound of formal systems, in spite of many conflicts, tensions and waste, continuing to be operational and producing relevant results. On the other hand, there are plenty of instances where voluntary organisations, cooperatives, professional associations, Rotary clubs and other service organisations fail, dissolve or end up being ritualistic in nature. Involving members in activities and increasing their participation and even getting them to attend meetings have always been a problem with such organisations.

Establishing dialogues in the social system is also another task. In political parties, educational institutions and professional associations cliques are formed. The cliques undermine legitimate authority and task coherence. They flood the system with processes of dissension, fragmentation and disruption. Taking decisions in such organisations becomes either a pseudocompromise or an empty statement of policy or agreement which is not implemented. Faced with a pseudocompromise or an imposed agreement, members have strong reservations and tend to activate processes to undermine the implementation of decisions.

The abilities to manage differences, deal with the primary task, and invest in the wellbeing of the organisation rarely emerge in these voluntary organisations. Every year, in the name of democracy, elections are held to posts for running these organisations. Having elected a new team or re-elected the old one, the membership sits back to let those elected perform the tasks. Members then evaluate performance and find faults. Their roles as representatives are limited to stating how they could do the job better and how inefficient the current role holders are. In many of these associations, as in political parties, even elections are suspended for years together. Ad hoc committees are set up. People are appointed, and the executive body remains an ad hoc assemblage of people chosen to balance the power equation.

In the emerging society and formal work organisations discovering and experimenting with new modalities of relationships is very significant. The Indian social design made no distinction between primary and secondary systems. It located the individual in a series of concentric systems from family to caste and finally to the endogamous community. There was no base for the development of differentiated qualities of relatedness between the primary and the secondary systems. Without discovering this differentiation and designing role taking processes, India cannot develop negotiable, effective and optimally performing formal systems. Organisations as well as every kind of public space would continue to be contaminated with the dynamics of personalisation and familial processes and indifference to collectivity spaces.

Extension of Internship Period

The Second World War brought the second revolution in technology. Technology moved from the mechanical base to a more integrated and complex base of continuous processes. This development introduced new fields of application of technology. An entire field of space, atomic energy and computer managed robots opened up. All these had a major impact on the nature and meaning of education. Earlier in India, education was largely service-oriented. For a large number of people education meant merely advanced literacy. It implied acculturation in the continental ethos (Garg and Parikh 1976). It had also been a source of emancipation and enlightenment. With the developmental thrust, education opened up not only opportunities for employment, but also became potential investment for the creative application of talents and individualistic achievement.

The widened scope of education encouraged students to go beyond graduation to the doctorate level. It made foreign education more desirable and consequently both the cost of education as well as the time necessary to complete a satisfactory level of education increased. Independently, certain legal and reformist movements raised the age of marriage. As a result, there was an increase in the period of internship of young people—social and educational—before they could enter the adult society. Thus, for all practical purposes the period of adolescence was extended.

The current generation felt that they were physically, psychologically and socially mature by the time they obtained their first degree, but the academic community and society at large still treated them as children. This created a gap between actual and socially legitimised maturity. Hence there was no space planned in the educational system for their level of maturity. Young people sought space for acting from this maturity either by rebellion or indulgence in what they believed were adult activities.

A number of individuals reported how families made educational achievement their sole goal and how they restrained their children from participating in the wider life space. It was only later that they realised how this exclusive focus, linked with the extension of the internship period, deprived them from realising their maturity. It denied them opportunities and/or time and space to recognise and come to terms with their fragile, raw, intense and adolescent emotions. They remained immature in terms of deeper and effective interpersonal relationships.

Indian society responded to these changes by allocating in its collective fantasy the role of *brahmachari* to the students. Having done this, the elite forgot that the *brahmacharis* of the traditional agrarian society lived in isolated and well-protected social communities. These communities were modelled on familial patterns and processes, within which a well-established infrastructure supported the individual in the role of *brahmachari*. Modern India's failure in this sphere is another example of introducing merely newer forms without making any effort to introduce a relevant infrastructure to help individuals make their new roles. Confronted with the prolonged period of internship and the assignment of an unrealistic role, young men and women coped with their adolescent impulses and the felt maturity by developing their own culture for exploring and experimenting. They were then blamed for moral degeneration.

This, then, became another road to cross in the process of transitioning from an agrarian to a technological society. Most individuals were caught up with the issues of boundary and maturity. What they felt at one level about themselves they doubted at the other. They were simultaneously confident and anxious. They doubted their being and fantasised their becoming. They were unable to internalise their success and trust their strength. Self-evaluation and self-affirmation were suspect, while external affirmation and evaluation were sought. Thus they were caught on a seesaw of assertion and doubt, unable to put in continuous efforts for a sustained involvement with tasks. They learnt to cherish short-term tasks where the results of their efforts could be assessed quickly. This undermined the quality of commitment and involvement they could bring to the self and system.

The Nature of Work Design

Another input of the technological thrust was the introduction of a work setting where people worked in groups. The design of the new work dimension was the interdependence of each element of roles and functions. The nature of new work demanded interlinkages, co-operation with others, understanding constraints, sharing responsibilities, and accepting the fact that one's contribution to the product was partial. These characteristics contrasted with the work design of the agrarian society where simplicity and sequentiality of the technology meant that individuals in isolation could complete the entire work. Major tasks requiring collectivity were designed around rituals and linked with festivals and agrarian seasons.

Thanks to the modern, formal work organisations, peer groups and collegiate systems appeared for the first time on the Indian work scene. These demanded that individuals learn to work in groups, exercise and influence task authority over each other, and manage similarities and differences. It required that individuals remain related to the task and system reality. It meant establishing face-to-face relationships with peers, and accepting their evaluation without feeling condemned.

Parents of those individuals who grew up when the technological thrust was just making its appearance needed only to accommodate themselves to the forms of technology and learn to handle the techniques and machines. The profession remained a work role and did not affect the core of personal identity which had meaning. They continued to derive their meaning from their social relatedness.

Once again Indian society adopted the same coping strategy. The new form was accepted but the need to redesign the process of transaction at the peer level was ignored. Most individuals continued to operate through their seniors. Face-to-face relationships were governed by sophisticated and socially acceptable behaviour. Behind this behaviour there was ruthless aggression. Healthy, mutually respecting and trusting relationships at task levels were difficult. Individuals in organisations often formed informal social groups on a parochial basis.

As participants explored and processed their experiences of growing up, it became clear that they lived a divided life between these worlds: a cognitive world rooted in the West, an emotional familial world rooted in traditional Indian society, and an action world which was not rooted anywhere. In their emotional world there always was a conflict between the shoulds of agrarian society and their feelings that arose in response to the changes. They oscillated between the desire to take risks and the need for dependency and control, between submission and autonomy, between assertion and passivity. They had to accept the monopoly of the role and the desire to break through it. They could not trust anybody, nor share, understand, explore, and process the surging emotions of their being. It was a lonely existence. The family, work organisation as well as society had failed to create spaces or systems with new social infrastructures. Preoccupation with the management of their complex life space left little time and space for parents to share and explore the changing context of society. This further impelled the current generation towards the socially desirable path.

What kind of societal field emerged for the current generation through the blending of their cognitive, emotive and action worlds described above? Our explorations with them led us to identify the following characteristics.

The societal field was an amorphous but dynamic collage—a kaleidoscopic mix, where elements of two distinct ethos acquired rhythm and ended up as counterpoints. The significant people in the societal field were anchored in the agrarian ethos and its accompanying network of attitudes. They accepted new forms but ignored or rejected the processes congruent with them. Thus the world of the current generation was characterised by ambivalence. In cognitive terms they held Western, liberal and post-Enlightenment values and beliefs and had faith in the techno-economic, techno-scientific and eco-political vectors of society. For these they had rational appreciation. They also had strong reservations regarding the social milieu of the West. Their roots in the Indian ethos and cognitive perspectives had been cut, and they negatively appraised the operative reality of India.

Individuals, thus, were faced with a conflict. Although they sensed existentially the subtle and indistinct ethos underlying the scientific and technological development, they could not clearly state it. They experienced a demand for action in the context of this ethos from within. The environment and its power elite, located in the agrarian shoulds, displayed a willingness to accept new forms, while discouraging new processes. This allowed very little scope for action from the ethos of scientific and technological development. These individuals, then, chose the soft options of imitating Western modes of social behaviour—consumer display of an acquisitive and extravagant lifestyle.

The tendency of society to treat science and technology as a means and an instrument of material civilisation, coupled with the failure of the intellectuals to articulate the concept of man and the collectivity inherent in the new science and technology, has left the current generation in confusion. They experimented with the ethos of science and technology. The societal field was denuded of forces which could challenge the individuals to fashion a new psycho-cultural identity. The imported ideals lost their credibility as the very preachers of those ideals stood exposed as hypocrites in the post-Independence era by doing everything to prevent the ideals from being realised. The cause and idealism of national freedom and participation in nation building and managing natural catastrophes, which provided significant opportunities for identification and fashioning of cultural identity, were no longer operative. India was already politically independent. Whenever a natural catastrophe occurred, the government monopolised all efforts to manage it. It appealed only for money and discouraged direct and voluntary involvement of people.

The new goal sets lost their potency because of unemployment or underemployment. Except for the challenge to excel in academic performance, there were no physical, intellectual, social or moral challenges which could mobilise individuals to fashion their identities. Even the challenge of academic performance, coming as it did from an education system which was devoid of idealism and values, did not inspire the student collectivity. Education remained largely techno-informative in nature, devoid of perspective building and intellectual challenges. Thus, there were neither forces nor support systems which could galvanise them to fashion their identity to contribute either to nation building or create belonging systems.

The current generation is perceived by the larger society as goalless and rootless. It also appears resentful and ambiguous in its attitude. Society's attempts to understand the current generation are hampered by the centuries-old traditional agrarian perspective. Likewise, the amorphous feelings of the current generation prevent it from understanding traditional society and its people. The failure of both to address themselves to the inevitable transition—the need for changes in the processes of agrarian social values and for reconstitution of the ethos to suit the current era—is the source of poignancy and the increasing disenchantment with each other and systems.

This failure has left the current generation to carry unaided the burden of its emotional world, which is a cauldron simmering with the guilt of belonging, unprocessed emotions and impulses of adolescence, reactive feelings, and genuine questions about the nature of reality. The systematic, and almost deliberate, neglect of the emotional world sows the seeds of scepticism, self-orientation, and disowning responsibility for replenishing the system.

What happens to the current generation when a lack of idealism is juxtaposed with the overfocalised and overcrystallised aspirations of material success and wellbeing, egoistic goal sets with diffused horizons and an uncertain future, and extended internship with the absence of socio-emotional proceedings?

The dynamics of this setting of growth has subjected them to two contrary and almost contradictory forces. The individuals were left in the nowhere land of no responses or wrong responses. They lived in the nowhere land of becoming without being. They struggled in a world of achievement without affiliation, and in the world of affiliation without achievement. They acted without feelings and felt without action. They floated into a shadowy world.

Our attempts to listen to and emphathise with the current generation in the settings of their growth have led us to identify some of the distinct patterns of the agrarian and technological ethos which, in a tug-of-war fashion, create the tortuous and almost annihiliating process of becoming. The impact of this transition is felt by individuals, formal work organisations and all aspects of structures, systems and society. In order to discover perspectives, values and action alternatives, it becomes essential to identify and understand the elements of the two distinct ethos. The next two chapters explore the elements of each ethos and their implications for Indian society and its people.

Four

The Two Competing Ethos: Emergence of the Culture of Transience

The social design of the Indian agrarian society embedded itself in the emotive maps of people. For centuries stability, continuity and consistency held society in its slow and steady unfolding. Over a period of time the unintended consequences of this process started to accumulate and to fray the social fabric and the social design. Morbidity began to set in. And what once had been a dynamic and vibrating society began to show signs of decay and degeneration, leaving itself open to influences from outside.

Across the geographical borders the West was preparing to make its impact on Indian society. Beginning with the first formal contact with the West, a process began which created upheaval and confusion in the Indian identity, culture, society and individual. The processes of intrusion, disturbance and dislocation accompanied this culture of transition and transience.

European culture and civilisation began to make inroads into Indian culture and society around the 1800s. By this time the Indian religio-philosophic ethos had ceased to be a dynamic process of social design. It had become frozen into agrarian structures, social rituals and customs; its dynamism had become a memory and a subject of scholarly study. It was a society drained of its vital energy; a society whose dynamic ethos had shrunk and which had lost its initiative and become preoccupied with the preservation of its once viable forms of social transactions. It was geared to regulate and maintain the framework of a structure which was once viable. Social processes geared to change, innovation and growth were so diffused as to leave the social design inert. It had lost the quality of self-regeneration.

This was the society which we have described as the society of role absolutism. It was left with only two functioning indigenous systems—the interpersonal and the economical system. Political authority and education were already in the hands of an alien power. Though the religious system was left intact. However, it was an isolated system in which the individual sought refuge from the stress of living. The economic and the interpersonal systems, at the family, caste or community level, were interlinked to the point of becoming fused into one. This role-centred agrarian society of India emerged from the collapse of the old society and the process got accelerated through the encounter with the Western ethos. Security came from entrenching oneself in the absolutism of the role processes. It also softened the impact of the alien ethos.

Two hundred years later after the entry, encounter, thrust, stabilisation and overthrow of the Western ethos, Indian culture and society were left with the legacy of the technological ethos. Nevertheless, the tenacity of the social design had left the societal familial process intact. As such, for the generation reaching adulthood around the 1950s, the societal field of growth was largely stable and displayed negligible variance over time and space. It did not contain many drastic social, relational discontinuities. This generation, here to fore, internalised the agrarian ethos to form the core of their role identity. They had been exposed to some aspects of the technological ethos which they had either rejected or sidestepped; they retained a strong agrarian mode of interpersonal relations and organisational processes. The elements of the technological ethos, learnt cognitively, became only a part of occupational and economic activity. These was not integrated into the network of attitudes to live by. These individuals were merely technicians not technologists.

The quality of this role-centred agrarian society can be discerned in some of the statements of the participants.

> You take away my family, its status in the society and I will be nobody; I am merely a continuum of the family unit, and in your words a mere extension of the family into the society; I am expected to live for the family. My life is mortgaged to the inescapable dreams of my father and my mother who would have liked to see me become a doctor but here I am. And now I am supposed to marry the girl my mother has chosen.

If we put together a whole set of such statements there can be but one inference. The transactional ethos of the role-centred agrarian society demanded surrender of the individual and his self to the family, caste, or what can be simply stated as the primary system of belonging. His temperament and character were merely a reflection of the family *samskaras*, and his success an expression of family investment. His entire existence was restricted to his social identity. This emphasis created a strong ethos of role identity and role relations. Most of the processes of individual existence excluded consideration of the self. Role-boundedness was cherished and praised. Anybody who attempted to free himself from role boundaries and sought to bring his own self into his behaviour was either rejected or made an outcast. The generation of the 1950s internalised and accepted three basic anchors and boundaries of the transactional agrarian ethos. These were: (*i*) the nature of social identity and the processes that contain it, (*ii*) construction of role reality and the processes that sustain it, and (*iii*) the work design and processes integrating the earlier two.

Social Identity and its Processes

The most significant aspects of Indian agrarian society were the rigidly defined and bounded systems of exclusion and inclusion defining an individual's belonging. The existence of clearly defined castes and their multiple subcastes, closely knit with their own myths and endogamous marriages, support the above statement. Name giving practices also provide evidence for it. In many parts of India the individual carried the totality of his social identity as part of his name. The name included the coordinates of its social role identities, such as the geographical location, the name of the household, the name of the father, the name of the *gotra*, the title of the caste, and his own name. The number of coordinates included vary from region to region, but a minimum of two were always included. The individual was firmly embedded in the context of his social identity. While introducing himself to others he first announced his social identity. The best example is the *Pravara* and *abhivadane* that an individual is supposed to articulate in introducing himself on ritual occasions in South India.

Housing arrangements in villages and towns further substantiate the claim. In keeping with the emphasis on social identity, the agrarian ethos evolved a pattern of cohesive social living. Homogeneous groups lived in compact neighbourhoods, where almost year-round festive and ritual occasions underlining the togetherness of social living were developed.

In spite of the rigidity in definitions of social identity, it is interesting to note that individuals in the agrarian society enjoyed dual membership—in the socio-psychological community and the socio-temporal community. The socio-psychological community was made up of blood relations and extended kinships. This determined the individuals psycho-cultural identity and provided psychological security and shared myths and rituals. The population of the unit village or a cluster of villages constituted another system of membership, i.e., in the socio-temporal community which was interdependent in many ways. It formulated the secondary norms of social conduct. Through governance by consensus, it freed him from the rigid hierarchy of the socio-psychological community.

At the dawn of the twentieth century the fine balance between these two memberships was disturbed. In most cases membership of the socio-temporal community became dormant, in fact inert. This is why social actions such as taking responsibility for common spaces came to be neglected. Looking to the wellbeing of the social-temporal community, in which all members should have participated, was left to the philanthropic initiative of a few individuals. Examples where the whole environment, other than the inner sanctum of homes, is treated as a garbage dump exist across the country.

The data from the groups suggests that the following consequences ensued for the individual from the centrality of social identity.

Socio-psychological Consequences

The agrarian society fostered closeness, dependency, cohesiveness and role rigidity. It also nurtured parochialism with the socio-psychological community and encouraged distance from the socio-temporal community. This is evident in today's society as well. It provided psychological security in relation to the system but created fear and anxiety in relating to others. All relations in this definition of social identity were given, codified and involuntary. Building of voluntary relationships was inhibited. The relations beyond the family could only be symbolic extensions of primary relations and were governed by the same underlying attitudes.

It was a very stable system. It charted out all duties and all the permitted modes of meeting life situations for the entire life of an individual. In the manifest social fabric there was no scope for conflict unless someone negated his role. The intrapsychic conflicts resulting from bringing the self into the role was an individual's burden. Living by taking a role was prescribed and making a role was proscribed. The merit of an individual lay neither in innovation nor in performance, but in conformity, loyalty and continuity of the traditions. Expression of self-assertion of any kind, except the role syntonic, was taboo.

The boundary of social identity tended to limit aspirations, change and mobility. Living for the role and the family was considered positive behaviour. It denied the individual a participant and representative role for himself in the affairs of the family and the group. He was a mere doer. The management of the system was left to the occupants of the role designed for it. The individual had to conform. Convictions arrived at through personal experiences in life could not be acted upon in the system. If he acted upon them, he was treated as a social criminal like Chaitanya. If he persisted he was deified as a saint, and in time rendered ineffective in bringing about any change. His ideas were idealised but not practised, as only a man of Chaitanya's stature, in terms of development and stature, could act on his beliefs.

The simultaneous dual membership of the socio-psychological and socio-temporal communities left the issue of commitment unresolved. There was no touchstone for choice of action and/or conviction. If an individual was caught between competing demands of the two communities, he had to choose between escaping the stress by non-action, making a hard decision and choosing the demands of one community over another, acting expediently to serve his vested interests, or surrendering to pressure.

Consequences Relating to Work

The structure of work was significantly affected by role absolutes. It became a mere replication. Successive generations used the same tools, the same materials and almost the same types of skills to reproduce the same product. Innovation was not prized. Work remained routine. It meant taking a role and was not an expression of the self. It is this attitude to work perhaps that prevented the occurrence of technological breakthroughs and inhibited creativity in the material spheres of life. It is this attitude to work that, even today, prevents individuals in technological systems from responding on their own to the demands of new tasks.

Consequences for the Individual

The whole thrust of the agrarian society and its anchoring of the meaning of existence in social identity was to create forces of stability. It constructed a reality for each individual and this reality was given from the outside. The personal–experiential reality was disowned. This is why one of the most central conflicts of the Indian identity revolves around the choice between commitment to the self and commitment to others or the system.

Earlier, individuals had the option to commit themselves to the role and disown the self, or commit to the self, disown the given role and live by responding anew to each situation. They could also integrate the two commitments and transcend the limitations of choices. In the myths and epics of Indian culture all three models are available. Rama, Bhishma, Yudhishthira exemplify commitment to the role; Dhruva, Prahalad and Krishna, to the self; and the Sapta Rishis and the sages exemplify the third alternative.

Agrarian society demolished the third alternative. An individual could choose commitment to the self only on the pain of becoming an outcast. Arjuna, in the *Mahabharata*, represents this dilemma of the Indian identity by asking the question, 'What is *kartavya* and *akartavya*?'

Consequences for Nation Building in Current Times

India was never a nation in the political sense; it was primarily a cultural entity. The very concept of nation was the product of exposure to the Western ethos. Since it was in the context of an agrarian ethos that the movement for nation building was initiated, let us examine its implications.

Restricting the meanings of the self to social identity and consequently to role boundaries worked against the inclusion of the self in the wider and broader social reality. The individual learnt to restrict himself to the concrete and immediately visible boundaries of social identities. Issues of higher and abstract-level systems of belonging were irrelevant. The history of India during the past nine centuries displays evidence of the destructive impact of this aspect of the Indian agrarian society. Small ethnic kingdoms, though repeatedly warred against by Islamic invaders and the British, responded typically with, 'Let them come to our borders and we will see.'

Time and again calls for united efforts against the aliens failed to move the people. Most people could only extend themselves to their own borders. If stories are to be believed, many kings and communities watched passively, and sometimes gleefully, the destruction of their neighbours. Even the efforts to unite the people during the First War of Independence (1857) collapsed because the masses could not grasp the concept of a nation. Once again, the current disturbances in Assam, Punjab, Kashmir and other states reflect not only the parochialism of the people but their failure to respond to an abstract system of belonging.

How were the consequences of the centrality of social identity reflected in the personal world of the current generation? How did they handle them?

Most members of the current generation perceived themselves as victims of role absolutism. The social structure and seniors demanded performance from them within the prescribed role boundaries. They did not give legitimacy to the Being of the current generation. Neither did they affirm them as individuals in their own right. Even though they felt oppressed, the individuals had to continue to relate to the social system which alone gave them economic and social meaning. Holding on to the system became a permanent modality. Attitudes of ambivalence became central in relationships. The source of opression lay in the enforced values of self-denial, sacrifice, detachment, surrender, passivity and non-action. They were doubtful about these values but did not know how to state the new values they sensed in their own living process. Any experimentation in those values and behaviour was stigmatised, so in manifest behaviour they surrendered themselves in order to obey and conform. If we have understood them correctly, their obedience, conformity and loyalty were primarily reflex actions. They were not self-chosen commitments.

Family: The Core System

The family is the primary institution where the individual's identity is fashioned. Let us see how the family operated in agrarian India and what were its structure and processes. The current generation has much to say about its experiences with the family. Itself caught emotively in the traditional role absolutism, the family had extended expectations of relationships. At the same time, it was sensitive to the emerging cognitive map of the environment in which the children grew up. As such, it often gave and conveyed contradictory messages. In envisaging and preparing the children for the future, parents often ignored feelings and experiences. In essence, over a period of time they lost the emotive links with the children.

The Indian family seems to be symbiotic in nature. It makes separation, distance and differentiation difficult, if not impossible. The current generation, while discussing the nature of its interpersonal relationships, stated how all its attempts to establish intimate relationships with outsiders led to unreasonable expectations of total inclusion. They had not learnt to receive a 'No' to even a small request. They felt let down. They were desperate to be owned constantly and continuously. To them every act was either a proof of love, loyalty and devotion, or of rejection and desertion. The Indian family fosters getting things done through personal pressures and emotional appeals. The merits of the issue do not matter.

Being extended, the Indian family of the agrarian society had many members. The structure of the family restricted each individual to a single role towards each of the other members and made all relations unidimensional. Relationships were permanent and in this the totality of the individual was lost. To a father, a son was a son for ever. Redefining and redesigning relationships was not allowed; thus they became stagnant and went through phases of closeness and withdrawal. The choice was either to retain the relationship in the same mould or give it up. Modification, review and redefinition towards an adult relationship was denied. Members of the younger generation chafed under this rigidity. They often asked:

> When will I be free from the son's role? Can I ever have a free and fair dialogue and personal communication with the adults of my family? Will economic freedom provide me the means of taking a stand and allow me to negotiate with them? Do I have to learn to behave like my own seniors who relate to the family members only in role acts and rarely share deep personal feelings? How can I avoid being in the middle, listening to my father's resentment against my mother and my mother's resentment against my father?

Why can these two not have a heart-to-heart talk and settle the issues themselves?

The Indian family fostered a rigid role hierarchy in which one individual member became the only significant person. The management of all relations in the role hierarchy of the family was controlled by this individual. All relationships were not only unidimensional, they were also characteristically superior–subordinate. No concept of equality existed. The superior–subordinate dimension is so ingrained in the society of role identities that equality of selves can never be achieved, and thus peer culture cannot be established. We see the failure of real participative processes in the Indian organisations as inherent in this social and role identity concept. Participative decision-making becomes the most difficult function in social systems in India. Unless this basic characteristic is altered, processes of formal organisations of secondary systems can become vitiated.

The extended family meant the distribution of emotions among a wider set of people. There was always somebody to love, somebody to hate, and still others to be jealous of. There was always somebody to play and laugh with. If reactive feelings mounted and frustrations piled up from the unidimensional expectations of the role, there was always an opportunity to seek a surrogate for appeasement. Thus emotions never became intense. They remained raw. Pent up feelings always found an outlet, making for catharis rather than discovering mature processes of relating with others.

The dynamism of the primary family gave no opportunity for proactive feelings to emerge, Individuals remained caught-up in reactive feelings which were discharged in immature ways. Emotionally, the child rarely became an adult. Immediate gratification, as discharges of tension in 'now-or-never' and 'all or none' modality, continued to control their actions. Gradually they learnt to acquire role appropriate feelings for most situations. Expression of all proactive feelings was considered a taboo. Their emotions did not get fully socialised.

Indian agrarian society discouraged moving away into the peer culture for the purpose of voluntary redefinition of one's belonging. A stage in the life-cycle during the classical period of Indian society, where movement away from the family was encouraged by sending the child to a teacher and letting him acquire a secondary identity, has been wiped out. It has been reduced to a ritual today. The child does not leave the home psychologically.

Obviously, such a system discouraged change in its processes. Change was seen leading to chaos. The thrust of socialisation and its medium, the family, was to generate processes which would overemphasise identification with primary roles. The individual's life was conceived as unfolding entirely within the matrix of family roles. The processes of the family system had been geared to overdetermine the quality of introjects and the nature of internalisations. In a very subtle way the agrarian ethos had struck upon the most effective process for perpetuating itself and for stabilising the rigid social structure of the family. The agrarian ethos conceived no change in social conditions, and promoted the belief that by the end of primary socialisation a youth would have acquired values, beliefs, cognitive processes, and modes of meeting life situations which would serve him for the rest of his life. The primary socialisation was the sole process of becoming. This belief has been transferred into the modern management systems where managements recognise obsolescence of technology, but rarely the obsolescence of skills and roles. That is why most organisations design training systems focusing on skills. For trainees training programmes are very often perceived as holidays or as punishment.

The Attitude of Obligation

Most individuals frequently used 'obligation' as a stock word to describe interpersonal relations. This was expressed in remarks such as, 'Parents have sacrificed a lot to do their duty towards us. Duty must be returned by duty, and sacrifice should be reciprocated by sacrifice. None of my actions should hurt the feelings of my parents.' Most individuals reported feeling guilty, anxious, helpless and tense when their resentment and ambivalence towards their parents were unwittingly expressed by them. Repeatedly, the phrase 'my obligation' occurred whenever the issue of choice between acting for the self or the family confronted the individual.

Authority: Legitimate and Non-questionable

Authority, residing in the parents, was considered legitimate. It could not be questioned nor its prerogatives challenged. Authority could be dealt with through personalisation, emotional appeals, pressures and finally through cheating. They pacified their consciences by saying 'what they do not know, does not hurt them.' But it did create guilt and anxiety. A majority of the young had to face guilt and anxiety after entering college because the peer culture pushed them into behaviour which was not acceptable by the family. A two-faced identity, one for home and one for college, was developed and they seesawed between guilt and bravado. But it also confirmed their manipulative stance, if not in every sphere of life, at least in the secondary systems. Conspiring against the authority of the secondary system became a well-established pattern.

The role-centred agrarian society of India appeared to have destroyed the earlier tradition of individuals locating the authority in the self. The agrarian ethos, like the Judaeo-Christian ethos, located the authority outside the self, either in institutions or in significant people in the community. It failed to generate processes to resolve the all-powerful image of authority outside the self which became that of a benign despot. The alien control by politicians and economic power reinforced the image, and authority became *mai-baap* (the mother and father). As such, it became a monolith of nurturance and control. People iearned that to be nurtured they must surrender absolutely to the control of authority.

Individuals internalising this concept of authority were left without processes to negotiate with it. The very idea of standing up to authority generated mortal fear. Gandhi had attempted to regenerate the processes which allocated authority to the self and from which the individual could act to deal with the authority of the system. He evolved the strategy of satyagraha for the purpose. However, after his death the strategy degenerated into a technique of defiance. How to transact with the authority of the self and the system simultaneously and effectively is one more dilemma for the Indian culture and identity. Processes which allocate authority to the self, and define modes of using them to contain the tyranny or of authority the system, have to be rediscovered and relearnt. The current modes of dealing with authority are dysfunctional. They do not contribute to the development of either a psycho-cultural identity or a healthy society.

Role-boundedness

You are only a role in the total system. Your safety lies in doing the role-acts whether you are motivated or not. Restrict yourself to

> the domain of your role. Do not concern yourself with the issues of the total system because that is somebody else's role. Do not poke your nose in matters that do not concern you. Do not step on somebody else's toes.

The message given to the current generation was clear. It was desirable to perform the role without demanding a share in the ownership in the system. 'The system owned you because it gave you identity. You owe yourself to it.' This was one set of attitudes.

The following quotation expresses the second set of attitudes cultivated by the agrarian ethos:

> Never talk of feelings. Hide them. Do not let others discover yourself. Keep a distance. Do not trust others. There is always a possibility of betrayal or ridicule. Build a closeness around doing things together, hanging together, to fill the loneliness in action and to remove boredom. But do not build strong emotional relationships. Be wary of real closeness. In brief, remain faceless towards the majority of role relationships in the secondary system.

This attitude revealed the emphasis laid by the agrarian ethos on being only a role and performing appropriate actions within it. It left no space for the individual to bring his or her psychological being into the relationships. Thus they felt utterly helpless. Role-boundedness and withholding of the psychological being in transactions led to the preoccupation: People think I am mad? Everybody knows it. Why should I stick my neck out? They perceived that hypocrisy was the hallmark of culture. They became sceptics. This perception further reinforced the withholding of their psychological being from transactions.

Individuals depicted their psychological being (self) in a variety of imagery. It was pure, sacred but fragile. It was ugly, intense and strong. It was invested with all the attributes they were afraid of. For example, quite a few individuals attributed to their psychological being the curse of Bhasmasur. They believed that if they really brought it out and touched somebody with it he would be destroyed. In the process the real experience of the psychological being was masked by these inferred attributions from transactional experiences. The process, in turn, reinforced the inability to initiate and act from the self.

The dual process of role-boundedness and withholding of the psychological being created the context through which the individual internalised symbolic identities of role and self from myths and folktales, as described earlier.

Work as Duty

> Work to achieve academic excellence. It is your duty. Parents expect it, so do teachers and employers. Parents sacrifice for your achievement. So study what is assigned. Work only when assignments have been given. Study only the directly relevant. Complain about the workload, but do it. Achievement is more important than learning. If you can achieve with minimum effort and by beating the system, do it.

These and similar exhortations suggest that role-boundedness banned involvement of the individual in the system. It also blocked genuine investment in learning and self-development. The current generation learnt to treat achievement as 'other-directed' and arising out of reactive feelings. No wonder today's generation displays instability and inconsistency in achievement.

The explorations also indicated that success and achievement, even when frequent and repeated, did not provide satisfaction. Success was not internalised. The need to be reassured about their adequacy continued to persist. Success and achievement did not seem to generate security, a sense of positive self-worth, or an ability to appraise realistically their limitations. Thus, even the most successful and achieving persons continued to compete within the system by pushing out other individuals. They wanted to be the sole achievers in the system and often destroyed others in the process.

Disowned Personhood

Under this label, the current generation subsumed all those feelings which it could not deal with. For example, its members felt they could not act autonomously. They perceived the family as overcontrolling, forcing dependence, denying maturity, and unaccepting of their growth. They felt excluded from decision-making about themselves and this left them imprisoned in the self. Many of them reported that they had foregone a scholarship to study abroad because the family did not wish it. Some of them had informed the family only after having changed a job, or given up a job to study further. Others reported that their only mode of dealing with the family was to present them with a fait accompli.

Discouragement of communication of feelings, denial of opportunities for exploring issues in an adult manner, condemnation of acting without advice or prior approval, reiteration of seniors' prerogatives and their message, 'you are still a kid, you do not know what is good for you', made individuals feel shackled. Their freedom was only to act in the role and not to feel and act.

In our presentation of this aspect of the transactional ethos of the role-centred agrarian society, we have restricted ourselves to the critical statements presented by the individuals in these groups. There were many an elaboration and corollaries to these main themes. This not only inhibited psychological involvement and investment in people and situations, but virtually became a taboo. It trained the individual into 'thinking into feelings rather than feeling into thinking.' It also reinforced the disowning of self-authority (autonomy). Individuals were trained to consider the shoulds and musts as their feelings. When confronted with their inner feelings in group settings, they felt ashamed of not acknowledging them. They reported that their seniors had instructed them 'not to feel as they do' and taught them to think in terms of 'should feelings' and 'must feelings'. Role-boundedness destroyed the ability to act as one's own representative in the system. It allowed the significant role holders in the system to deny dignity to the rest of the members.

Another impact of the processes of role-boundedness was to destroy the link between the formal space of transactions and the 'gallery'. It is not infrequent to find an individual making role-appropriate statements in the formal setting and then coming out onto the gallery and contradicting them. When confronted with their inconsistency, the individual invariably would say, 'These things cannot be said there.'

Maturity

At the manifest level of role behaviour individuals need three kinds of maturities.

1. Sexual maturity

2. Social maturity
3. Occupational maturity

The classical society of India had recognised that behavioural manifestations of maturity were not enough. Cultivation of attitudes congruent and coherent with the ethos was also necessary. Hence it planned two distinct periods of moratorium for the individual. The first was between the ages of thirteen and twenty-five. This was the period when an individual upper-caste child was placed in the *gurukul*. He acquired and perfected occupational skills and moved away from mere rote memorisation of the Vedas to the decoding of its knowledge through experiential processes. This prolonged period also achieved the goal of dissolving his primary dependence on the family. It promoted the introjection and internalisation of secondary systems as a basis for social living. It prepared the individual in the processes of role taking and in designing his role for adulthood. The emphasis was on linking the ethos with options and choices that the society permitted or which could be created by the individual in his own life space.

The second moratorium emerged when the individual was supposed to close his *grihastha's (karta)* role and hand over the charge of the household to his son. He spent his time in processing and facing up to his own being through a review of his life. Indian social design, then, clearly demarcated periods for review of the self and its relatedness with the world of transactions. Over the centuries, as the ethos and the social design became eroded and the agrarian moorings became more significant, these two phases of review and encounter with the self and system also vanished.

In the agrarian set-up three kinds of adult roles—sexual, social and occupational—were compressed into a short span of three to four years.

Sexual Biological Maturity: This is an inevitable aspect of growth. Most children between the ages of thirteen and nineteen achieve it and become capable of reproduction.

Occupational Maturity: This too was acquired early. The technology of an agrarian society was basically simple and repetitive. A child growing up in an agrarian society needed only to learn the basic practices and processes of his father's occupation. Hence, an individual could and did acquire occupational skills through apprenticeship within the family. By the time he reached biological maturity he had also acquired occupational maturity.

Social Role Maturity: This converged with both sexual and occupational maturities. The processes of primary socialisation were potent enough to help the growing child successfully internalise the interpersonal role-orientations needed in society.

Thus, for all practical purposes, an individual could easily step into an adult role. Society had designed decentralised autonomous communities and lifestyles which had well-defined modes of meeting life situations. Society was stable and ensured that the internalisation of behaviour and attitudes would lead to the successful performance of an adult role.

With the collapse of classical society and its shrinkage into the agrarian role–absolutist society, the psychological moratorium phase was dropped. The individual was expected to walk into the adult role when the three maturities converged between the ages of thirteen and sixteen years. The development of the psychological component of these three maturities was not considered necessary. Thus, instead of letting individuals construct a self-role reality, agrarian society required only the construction of role reality. The three maturities converged within a time span of three years, in which time the individual was expected to acquire all the wherewithal of his role identity.

The rapid convergence process of the three kinds of maturities deprived the child of the opportunity of understanding the self. He was left with no time to deal with his experiential feelings and doubts about the self and the social system. He also had no time to integrate himself. A role, fully constructed, was offered and he slipped into it with ease. It was not unknown to prescribe early marriage for a youth who tended towards apathy, listlessness or wanderlust. It is interesting to recall the actual phrase in the vernacular, *khunte se bandhna* (tie the individual to the stake).

The situation described above developed over a long period of time. Up to around the 1950s, when the parents of the participants in our groups were growing up, this was a standard practice. Through their efforts, the Brahmo Samaj and Arya Samaj movements had been able to push the marriage age up to eighteen and above. It seemed that somewhere along the way Indian agrarian society went through a trauma and became overconcerned with the construction and maintenance of a role action society. In its zeal, it even stemmed the flow of creativity that the young man, through contemplation, could have brought back for the society's rejuvenation.

Society tolerated some lapses, inconsistencies, and occasional escapades as part of the process of growing up, but ultimately it shepherded the individual into a role. The agrarian ethos sought and devised ways of postponing feelings, both experiential and personal. It pushed the individual into the matrix of role acts. Procrastination here could have been dangerous as it would have provided an opportunity to feel into his being. This had to be avoided by enforcing the convergence of the three action maturities. Thus, the individual in an agrarian society moved into his role for life. He did everything for the role. Even his marriage was for the role. He did not see the girl he was to marry. He only looked at her social identity. Inherent in this was the belief that an individual is what his family is and, as such, not the person but her *samskaras* should be looked at. Often a girl was married off to an unsuitable boy because his *samskaras*, inherent in the social identity and the role act heritage of the family, were matched. However, changes in attitudes towards marriage began around the 1930s. For the aspiring family modern education became necessary. Occupational maturity was no more a matter of routine learning through induction or apprenticeship on the job. It became a matter of formal learning and training. For the first time, the three maturities began to diverge. The complete adult role jacket which a young man could slip into was no more available. He became sexually mature by the age of sixteen but not occupationally mature. He internalised the interrole system of duties and obligations. A halfway house came into existence. Sexual maturity was recognised by arranging an early marriage which granted him his social role. He then proceeded for higher studies since he was occupationally not mature. This process gave youth the time and opportunity to view the world. However, his role involvement was so strong that the adoption of the new Western ethos and technology remained largely restricted to the development of a work role and redesigning modes of social living at work. The home and the domain of women remained traditional.

By the time the youth of the 1930s attained parenthood the scene had changed even more. A longer internship was needed for occupational maturity and this extended internship was too expensive to get the young man married as well. The agrarian process of convergence, ensuring entry into the role, was completely disrupted. Social living was postponed for a period of six to eight years, making the phase of psychological moratorium once more available to youth. But society failed to recognise this. It continued to hold the role and its performance as the basic values. No action to build social and psychological systemic infrastructures, so as to utilise the phase of psychological moratorium, was designed in the education system. So what the current generation inherited was a period of 'as if' existence, wherein they were supposed to remain in a state of suspended animation and only relate in action to the set goals (Garg and Parikh 1976).

The failure of society to design the institutional infrastructure for effective use of this phase led to scepticism and a sense of being let down. Attributing hypocrisy to adult society and disengagement from the system began. The education period became a space for escapades and indulgences based on impulse. Intellectual discourse and involvement that characterised the Indian universities between 1900 and 1938 vanished by 1947. Students were left with no institutions devoted to the integration of identity. The mere performance of roles, with its meaninglessness and ambivalence, was revived. The current generation was caught in the external demands of role action and its norms while within they lived in a turbulent world. They did not know where to turn for resolution. The only option was to deny the internal reality and to appear overengaged in the external reality of performance.

Many participants bemoaned the hollowness of their maturity. They were acutely aware of having been denied the psychological component of their maturity. They felt cheated by the system. They thought they were socialised only to achieve the social status desired by their parents. They realised their lack of zest for living, their apathy, the inability to mobilise themselves, and the tendency to procrastinate and fulfil role obligations through the process of management by crises. They felt the 'as ifness' of existence during education. They also recognised that their anti-intellectualism, lack of interest in serious literature, and search for excitement in drugs, drinking and jam sessions were due to their incapacity to break the bondage of an 'as if' existence. They were incapable of creating institutions for integration during the phase of psychological moratorium.

Sporadically, smaller, well-contained groups did develop such institutions for themselves. These individuals found some integration and ended up falling in the category of 'heirs apparent'. These institutions were operative in private spaces for small groups. For example, in academic institutions of higher learning, such as the IITs (Indian Institutes of Technology) and the IIMs (Indian Institutes of Management) the 'bull session' has become a permanent feature. However, often it ends up being a space for opinionated individuals to argue, rather than being a reflective session and an opportunity for bringing their psychological beings together.

In our initial dialogue with the second and third waves of the current generation we explored the reasons for their failure to utilise this period of psychological moratorium for their own integrative processes. Their common response was, 'Where is the time? Moreover, neither the system nor one's peers encourage it. Those who try are ridiculed.' The system at best encourages the individuals to use this time to meet the demands of their social responsibilities. It also promotes social work activities. Most institutions do not even do this. They either leave the students to their own devices or see their obligations as being fulfilled by providing for extra-curricular activities.

The second period of psychological moratorium took place during one's middle age. Agrarian society once again ignored the fact that long periods of being in work and household roles could erode the meanings that an individual carried from his adolescence to adulthood. It operated with the belief that once socialised in the role ethos the individuals could continue with it for the rest of their lives. It made no efforts to design new regenerative infrastructural processes for this period.

The Concept of Work Role and its Processes

The ethos of the agrarian society, in its size, scale, design of work, and technology, retained the concept of a producer society. An individual's work could be done alone from others. A potter could carry out his production alone, as could a weaver, a fisherman, a carpenter, a blacksmith or, for that matter, a farmer. A farmer could spend his entire day in isolation, in his field. When help was required family members or the community joined in. Almost all the occupations were so designed that the individual worked with his materials and tools in isolation from others. In the agrarian society, then, the individual lived in a group but worked alone.

The psycho-cultural identity which encouraged relatedness with a wider universe of experience and living space during the classical era had shrunk to the limited routine of managing a subsistence level existence in agrarian society.

The participants spent a great deal of their time talking about the issues of work role and the nature of work organisations. For most of them work by itself was not important. What was valued was the status of the work organisation. The power and the package of perquisites were more important than the nature of work. It was strange to note that both the youth and the work organisation colluded at the entry point in creating the myth of challenges available in work. According to most participants the work they were given was repetitive and non-creative.

They also recognised that it was difficult for them to work in groups. They admitted to the difficulty of creating a team spirit. They said that very few of them were capable of sharing responsibilities and upholding their commitment to their share of the teamwork. They needed to be goaded into action. It was their experience that the Indian work organisation was characterised by non-delegation and non-sharing at the peer level. They had a tendency to portray themselves as being indispensable.

In the light of the above statements, it was not surprising to discover that to most individuals work was only a means of livelihood, not a part of the core meaning of being an individual in society. At best it was a means to achieve social status. The task and efficiency were not as important as power, status and control of resources in the work role. Work was just another aspect of role absolutism and, as such, constituted the social identity and not the psychocultural identity as it had in the classical era.

Our explorations of the concept of work and work role with the first wave of the current generation brought to light a whole universe relating to the role absolutism of agrarian society. Some of these participants were already in their late thirties and early forties. They were discovering an emptiness in themselves. Frequently, statements like the following were made:

> I have never taken leave to be with the family or gone on a holiday. I normally cash the leave travel concession. My wife accuses me of being married to the organisation. I leave early in the morning and return late. I bring work home. I cannot depend upon my subordinates. I have to check everything. Young men entering organisations may have knowledge and appear smart but in prac-

tice they are incompetent. They are theoretical. They want to change too much too fast.

They also talked of their success and how they have perfected a mode of success in the organisation.

The repeated themes were of hard work, routine work, compulsiveness, interrole and interdepartmental conflict, struggles for developing managerial and leadership styles, managing the power equations in the organisation, and dealing with the rigidities of superiors. Feelings of being discriminated against, being cheated in evaluation and bypassed in favour of less capable people were frequently voiced. Continuous effort and application, postponing satisfaction, personal needs and pleasures for work, demands of psychosomatic illnesses, and occasional catharsis through drinking parties and smutty jokes cropped up often in their explorations.

We attempted to sift the data in order to get at the role concept embedded in their experience of work and work organisations. At one level it reconfirmed all the themes identified in the description of the agrarian ethos. The following paragraphs illustrate the point.

It was obvious that there existed a dichotomy between designed and emergent organisation behaviour. The emergent work organisations were psychologically and emotionally based on well-conceived designs and modelled after the family. In the cognitive plan, its complex design and structure did not exist. The roles, their psychological aspects, and the role processes in most cases paralleled the roles individuals had played in their own family systems. Sometimes the parallel was frightening. A 'youngest son', in spite of his higher designation, acted in the role processes of the youngest son. This illustrated the coercive and complete socialisation in role-boundedness and absolutism in the lives of the first wave of the current generation.

The allocation of the authority to the self was completely absent from work organisations. In their interface with the superiors, even the role authority was shaken or denied. Managers very often quoted examples of their role authority being undermined by their superiors. The issue of authority, non-negotiable and rigid, remained unresolved. It not only blocked initiative and creativity, but also failed to invoke commitment, a sense of belonging and satisfaction.

The work role had denied them their *self*. But it had helped them acquire all the necessary symbols of having made it in life. They had delivered; in return, this had generated the socially desirable standard of living. In addition, they had capable wives and smart children who went to good schools and were good achievers. Awareness of these benefits made them hold on to their role, their modes of meeting life situations, and repeating their past modes endlessly.

Residual feelings of lack of communication, helplessness and loneliness all existed, but what was more important was the question, 'What next?' When asked what next, they realised that the meaning the work role had for them in the beginning of their career had weakened. With it they also lost hopes of conquering the world. The spirit was lost, and a mechanical quality had set in. Feelings of uncertainty, inadequacy and despondency had multiplied. As outsiders we recognised the peculiar desperation and entrenchment experienced by these individuals. They wanted to change their role and renew and reanchor themselves in life while still working in the same setting. However, they did not know how to do so. They were aware of the stress and could not manage a psychological shift in their orientation. While attempting anything new they were filled with apprehension. Hence staying put in the same mould seemed more desirable. Owning up their personhood was difficult at this late stage in life. Turning to religion, faith and the supernatural was easier than making efforts in the direction of rationality.

Men were not alone in experiencing the ennui of the middle age. A sizeable number of housewives from the first wave of the current generation who participated voiced similar feelings. They had lived a lonely life while their husbands were busy building their careers. Looking after their growing children and managing the interface of social living had kept them occupied earlier. However as children grew up, they were not needed as much as before and the management of social living became exhausting and boring. Their husbands' colleagues' wives were the only women with whom they had any relationships. These were not always open and enriching; in fact, they were mostly formal or casual relationships. The hierarchy of the organisations also governed these relationships.

Making the home beautiful, cultivating graces, tastes, and such other 'feminine' modes of expressing the self, and finding meaning in life were not sufficient anymore. Being visible in public by visiting clubs was not fulfilling. Surprisingly, many of these women at this juncture became interested in economic activities. Still others thought of engaging in social work. In most cases either these alternatives did not work out or became a source of conflict with the husbands and children.

Thus, a lack of recognition of role obsolescence and a neglect of the need to design an infrastructure for self-renewal by society created forces which kept the individual deeply entrenched in role absolutism. From our point of view, one of the major challenges of Indian culture today is to accept that meanings and orientations of role taking become exhausted. The same individual at different stages of life needs a space and an opportunity to reanchor and rededicate himself into a new kind of relatedness with society. The period of psychological moratorium in adult life has to be recognised and accommodated into the social design and culture if we are to move away from role absolutism.

The dynamicity of classical society and the stability and continuity of agrarian society had made its impact on the psyche of the Indian individual as well as society. Classical society was an ideal and a cherished and valued heritage while agrarian society was burdensome and restricting. It was seen as a source of captivity, subjugation and surrender, not only to the role absolutism of the family and society but also to the domain of an alien, Western ethos.

Even as the Indian psyche and society struggled with the interplay of elements of classical and agrarian societies, came the encounter with the technological ethos. This confronted Indian society with a double-edged sword. One edge was the hope to grow into a new order and society and build a new heritage, while the other edge confronted a model so alien that it created waste and despair. Both these processes were once again anchored in a comparative frame. India as a society began to gather its energy to hasten its growth. The pressure of growth and the constantly shifting horizons to farther and farther destinations created enormous social disarray. The social fabric, already stretched in multiple directions, began to fray. And like the earlier period Indian society and psyche were not prepared to design social, psychological and systemic infrastructures to sustain, maintain, foster and nurture the direction of growth.

Five

The Ethos of Technology

The use of technology is universal. Most cultures have acquired a symbiotic relationship with it. With reference to capitalist/post-capitalist industrial technology, some hold it in abeyance, keep it at a distance, view it with awe or suspicion, want it but cannot manage it. Essentially, in today's times technology and its infrastructural network have become synonymous with success, growth and development. It is the varying uses to which technology is put which differentiates the First World countries from those of the Third World. It is ownership of and control over technologies which distinguishes advanced and developed countries and cultures from their developing and underdeveloped counterparts.

Every country and culture at some point of time has experienced the major discontinuity from an agrarian base to a technological industrial base. This discontinuity has created unforeseen disturbances in existing social structures, processes and roles, modes of meeting life situations and interpersonal relationships. Societies had to undergo realignment and recalibration in processes to integrate the introduction and increasing complexities of technology.

The countries and cultures of the West soon realised the enormous powers of modern industrial technology. The religio-philosophic assumptions of the West about man, the collectivity, and the relationship between the two created attitudes, beliefs and values in response to the technological advances. These assumptions facilitated the countries and cultures to incorporate technology as an integral part of living and as an inevitable aspect of the growth of human civilisation.

There were many other cultures and countries and even continents whose religio-philosophic assumptions about man, the collectivity and the relationship between them continued to focus on other dimensions of living, social organisation and relationships. Another critical factor which provided momentum to the West was the availability of certain kinds of resources and freedom of opportunities. At the beginning of the seventeenth century the West subjugated geographical locations of minerals and fuels as resource generators for themselves. These resources were required for the development of modern technology. Both these processes made a sector of the world the initiator and contributor while leaving the remaining as a recipient and borrower. Once this pattern was set in motion and gained momentum, it created a powerful, affluent material culture which dictated its own values, structures, patterns and processes to reinforce the demands on the individual to act within the given roles and structures.

The countries and cultures which accepted this technology as a symbol of economic growth and development had not envisaged the force of its impact on existing social structures, systems and modes of relationships. The Indian society of the classical era, the social modes of the agrarian society, and the two relatively new, distinctly unique religio-philosophic traditions, that is, Islam and Christianity, responded to the introduction and impact of technology in an unpredictable manner. Centuries of absorption and assimilation of multiple aliens and their ethos had left Indian society vulnerable and in disarray. The healthy core had shrunk and built a protective wall. This was not sufficient, so that the nation lay open to aggression, encroachment, impingement, stigmatisation and accusation.

It is at this juncture that dialogues with the current generation began. The objective was to discover the elements of the technological ethos as experienced and responded to by them and why. What were the elements which were being ignored and why? The culture, country and its citizens—filled with the hope of a new tomorrow and the despair of centuries, aspirations for the future and the helplessness and decay of centuries—were ready to take the step into another century to create a new paradigm. What happened and why. Let us take a look.

Since the 1920s Anglo-American literature on the nature of science and technology and their relationship to society has grown. The focus of these writings has ranged from the philosophical work of Alfred North Whitehead to the descriptive writing of P.C. Wren. Futurologists have provided another dimension for anticipating changes in lifestyles and social structures. Science fiction has tried to present the shift in human psyche, and the human context of these writings provide a rich understanding of the transitions, emerging phenomena and their implications. Naturally, these did not explain the Indian phenomenon. From these writings therefore we could not arrive at a model which would reflect the nature of this ethos of technology and the shape in which it was emerging in India.

Perhaps we were not willing to draw up a universal model and accept it as the reality of India. We were committed to articulate the assumptions of the ethos of science and technology as sensed and experienced by the people living in our society. This created a major problem. Behavioural, attitudinal and cognitive data displayed by the current generation was confusing and stereotyped. Much of it only mirrored the manifest forms of the Western societies. To achieve our objective, we attempted to go beyond the manifest behaviour and to delineate the context of science and technology in our society.

Western technology was introduced in India as a means for economic advancement. It was associated with efficiency and productivity on the one hand, and machines and complex systems of management on the other. The focus was largely on the adoption of technology for production and managing its interface with society, particularly overcoming resistance to it. The choice of importing technology was accepted as rational, both economically and politically. Those who were keen to introduce the new technology were only concerned with its technological, economic and political implications. Technology was never viewed in the psycho-cultural and psycho-philosophical perspective. These perspectives were used as pegs on which to hang the failures of technology.

The participants in the study perceived the introduction of technology as an opportunity and to create new professional roles. It was a means to fulfil aspirations and claim membership in the universal society of the modern generation. They had no vision of their new roles. There were no indigenous prototypes on which the new roles could be modelled. Their own psycho-cultural identity and their identification with the continuity of the Indian ethos had already been eroded, as discussed earlier. As such, the current generation was very vulnerable. The planners of technological development had failed to build social and psychological infrastructures to provide direction and support. Altogether, a whole generation was left to find its own way. Its response was identical to that of the planners. They too became borrowers of the social and psychological modalities of the West. Modernisation was equated with Westernisation by them. However, in the relational field they were still hooked to the emotive patterns of the agrarian ethos.

In the early 1950s the focus of models of growth shifted from Europe to the United States of America. In this shift, the current generation lost touch with the deeper philosophy and thought of Europe and was nourished on the techno-informative knowledge, behavioural dynamics and conspicuous consumption of American behavioural patterns. Knowledge became merely techno-informative and conspicuous consumption grew as an evidence of success. The current Westernised generation appeared as anchorless as before. They were superficially acquainted with the philosophy of existentialism. They were vaguely aware of the need to participate in social reality. They occasionally talked of nihilism and expressed the wish of opting out and creating something new. But on the whole the current generation was largely unaware of the philosophy, poetry and literature of the culture they were trying to introduce into Indian society and into themselves. Of course, there were a handful who were in touch with 'isms' of the West. From among these some took to Marxism, some to existentialism, and others to scientific realism. There were very few who attempted an eclectic perspective. It seems that being anchorless made them so anxious that they let themselves be proselytised rather than acculturated.

Thus, the current generation camouflaged its deeper anxieties and concerns with identity by adopting a collage of behaviour, action and jargon borrowed partly from the sophisticated continental culture, and partly from the brash American culture. This collage became their frame of reference. They felt anxious, guilty and confused, but did not have the courage of their convictions to pause, ponder and integrate. They conformed to the trendsetters wanting to be with the ingroup and doing the in-thing. We found that the current generation wanted the best of both worlds. They wanted to consolidate the secondary gains of Western culture but were afraid to respond to the primary commitment it demanded. They were exasperated by the commitment demanded by the agrarian ethos. They were, however, desperately holding on to the secondary gains of the agrarian ethos. Thus, the youth appeared uncommitted, self-engrossed, self-centred, and wedded to the notion of expediency.

The current generation attempted to acquire and activate the forms of an alien culture and behaviour. These were not anchored in relevant attitudes or supporting processes that sustained the legitimacy of these forms. This created morbidity. It disturbed the social as well as the individual structure of identity and role taking and in turn blocked internalisation. It promoted intellectualisation, and a tendency to hair-splitting arguments. It negated the identity of the self, whether national, social or personal. This led to scepticism.

Isolating and articulating the elements of the technological ethos was difficult. However, in our explorations with the current generation, the impact of the technological thrust came up again and again. Socially, in terms of relationships, and psychologically, in their emotive aspects, the individuals felt that the demands inherent in the technological ethos pushed them from a feeling of stability and certainty into a state of flux, ambiguity and transience. For the first time individuals were confronted with a multiplicity of behavioural forms, beliefs and values. This made them question what they had been brought up to believe. They found themselves aspiring for and being challenged by the new opportunities. Responding to this created conflicts with the prescriptive aspect of role structures and given relationships. It raised doubts and dilemmas about living for and with others. For the first time the individual was confronted with the option to choose between the self and a role. They had an opportunity to design their own lifestyles and ask questions of themselves. What gives meaning? Relations or performance? Affiliation or achievement? The choice of creating a relationship instead of merely living according to a given one brought the individuals to the edge of insecurity. They were caught in an 'as if' existence, where the forces demanding unfolding of the self and creation of a new reality were locked in a battle with the ethos internalised from the family system.

The 'as if' existence of the current generation and their state of being is analogous to the story of Sleeping Beauty. The story symbolically represents a call to surrender dependence and nostalgia for living in the past and to create a reality of one's own. Like Sleeping Beauty, the individuals in the second and the third waves of the current generation seemed to be awaiting the arrival of a Prince Charming who would give them the kiss of life and awaken them to their reality. In their search for roots they followed the trendsetters among peers or the media. Very often they waited for the significant others to define the new reality of their beings. To describe them thus is not to say that they did nothing. In their own manner they challenged the system by seeking conditions for working in groups and pushing for the legitimacy of interdependent roles. They demanded delegation of authority and responsibility at task-levels, and tried to create a functional and negotiable group culture.

As the technological ethos acquired more momentum and its inherent processes began to make demands, the individuals awoke to the changing realities of roles, relationships and the new interfaces with people and environment. The realisation and encounter demanded new action choices from them which triggered off and reinforced many intrapersonal and intrapsychic conflicts. The individuals suffered from guilt, dysfunctional anxiety, indecisiveness, tension and turmoil of a dual identity (Garg and Parikh 1976). They also described their poverty of convictions, loneliness, doubts of self-worth, issues of morality and self-image and the struggle of choosing from multiple goals. They encountered uncertainties and definitions about duty, gratitude and responsibility to parents, the anxiety of disappointing them, and fear of loosing their approval, support and affirmation. The individuals found themselves trying to settle the question, Should they let themselves be owned and valued by family and organisations or learn to own and value themselves? As we processed these interpersonal and intrapersonal conflicts and dilemmas with the current generations, at first glimpses and then gradually elements and characteristics of what may constitute the ethos of science and technology began to crystallise. The impact of the technological ethos reverberated through all aspects of social and work institutions, structures, roles and relationships. It generated concepts of a new society. In the following section we discuss these elements.

Meaning and Identity in Work

With the introduction of technology formal education became a critical constituent of work. Education became linked to occupation and with it economic and social status. With the universalisation of education it became inevitable that occupation and career paths would be dependent on performance and achievement. The individuals of the current generation had begun to sense that in the emergent new society their personal assets of intelligence, ability, drive, motivation and application would determine their socio-psychological location in the family as well as society. In fact, they sensed that they would have to create an identity for themselves around competence, success and professional status. They realised that their family anchored identity would offer some advantages initially during the process of transition from the agrarian society to the emergent technological society. As such, one of the significant constituents of the technological ethos in the area of work was job satisfaction, challenging work, need for innovation, creativity, enhancement of role, career path and professionalism.

This shift from the agrarian concept of work linked to caste and familial occupation to a technological concept of work linked to education began to enter the social setting. This was felt most strongly in the area of marriage. In terms of the individual's social worth, a shift was indicated in the marriage market. The parents of the prospective bride tended to give greater weightage to the professional degree and the work organisation of the groom than to the family background. The technological ethos envisaged large and mammoth organisations. These organisations designed formal structures with multiple levels of hierarchy, specialisation, technical and managerial competence. Increased heterogeneity meant performance appraisal. In the technological ethos, the individual within his life-time could rise through many levels of hierarchy and improve his work identity by constant application of the self. Consequently, he could also improve his social worth and economic status.

The technological ethos brought about a major shift in significance, individual as well as social. This was distinctly different from the preset conditions of social design of the agrarian ethos which restricted social and personal aspirations. Under the impact of the technological ethos, the rigidities of occupation linked to family, caste and community started to dissolve. The individuals were exposed to rising aspirations and expectations. They could choose the occupational reality they wanted to belong to. Very often many individuals disowned their belonging in the social system in which they were born. Geographical mobility further reinforced this process. As such, one of the critical and significant constituents of the technological ethos emphasised individualised effort and investment in work organisations. Increasingly, job satisfaction, career opportunities and professionalism acquired significance and meaning. This demanded from individuals persistence in effort, constant readiness to respond, and willingness to invest in themselves. These dimensions were recognised as the essential conditions of social and psychological movement from the agrarian ethos constituted by fixed moorings of role, social identity, lifestyle and the community context of membership.

This shift and movement of individuals from the traditional agrarian ethos to the emergent technological ethos created many socio-psychological consequences.

Emergence of New Definitions of Interpersonal Concepts

The shift toward work identity as the source of meaning created new interpersonal demands which created conflicts in the participants. The conflicts were inevitable because the definitions of interpersonal relationships and transactions given by the agrarian society and those suggested by the technological society did not converge. One of the critical constituents of the interpersonal transaction was the evolving meaning and definition of relationships. In the technological ethos a concept of a secondary system as distinct from the primary system emerged. The ascriptive role structure came under pressure as more and more individuals sought open-ended relationships.

New role relationships had to be negotiated with the same people over and over again to create new boundaries of roles and maintain the quality of relationships. Due to geographical, social and psychological mobility, the quality of relationships was continually changing. Individuals were pushed to reinvest themselves in their relatedness with others in order to retain relationships. Relationships could neither be taken for granted nor postulated with the same expectations. Redefinitions and redesigning were an inherent dimension of relationships.

Obedience, conformity and loyalty acquired new meanings. Obedience became a matter of a negotiative decision-making process to which individuals were committed. It could not be accepted as a mere right of authority or a demand from structure or hierarchy. It allowed individuals to bring their experience into social and work transactions. In fact, the superior–subordinate relationships, instead of being fixed forever as in the agrarian ethos, became mobile and potentially reversible. In the formal work organisations of the technological ethos, technical skills, quality of education and performance could lead to superiors being younger in age and experience. The older role holders could, despite their years of experience, be subordinates to younger superiors.

The constituents of the interpersonal relationships of the agrarian ethos acquired newer meanings in the emerging technological ethos. Conformity, like obedience, was necessitated by participative norms and goal settings. Individuals could negotiate, act as their own representatives, and thus influence the group norms to which they then conformed. This became the basis for participation in group processes and promoted a sense of equality and assumption of egalitarianism. The formal work organisations made responsible self-choice a matter of commitment rather than a duty. It promoted inclusion of the self in the authority structure of the group and created an organismic concept of 'we-ness' instead of the mythical concept of a work organisation being one happy family.

Similarly loyalty, instead of being a matter of role relationships and as such demanding emotive investment, was transformed into a rational commitment to the goals and the norms of the group. It was no more dependent on the quality of personal relationships, but was transformed into a rational agreement around issues which the group faced. In such a situation an individual could agree with another on one issue but disagree on other issues without being accused of disloyalty. Loyalty as such was to the tasks, organisation and the institution, not to one individual.

One of the critical and significant constituents of the technological ethos demanded redefinitions of basic concepts of life structures, systems and relationships. The assumption governing the emergence of new definitions was that individuals were representatives in a group. Groups therefore were voluntary. Individuals also related to much more than the system and this had to be taken into account. An implication of this assumption was that an individual, being more than his role, could relate at multiple levels simultaneously in the same system or with the group. This was quite different from the monorole construct of agrarian society.

Nature of Authority

The most substantial element of the technological ethos related to the redefinition of authority. The emergence of a representational aspect of the self in the work setting implied that experiential elements could be brought in for negotiation and decision-making. This rendered the authority transient and a matter of agreement. The inclusion of the self in the structure of authority of the group freed the individual, allowing him to question and censor the role authority. Individuals could negotiate for new goals, roles and norms. Failure of negotiations might reduce the individual to a minority, but he could retain his membership and work from within, hoping to convert the majority to his viewpoint by rational insistence on the issues facing the group. Or he could opt out if he so chose. Authority was no more a unilateral social process anchored in structure and hierarchy, but a reciprocal one with processes of participation.

Another critical constituent of the technological ethos, viz., self-reliance and the kind of dependence on other individuals, emerged as positive aspects. The technological ethos had no place for dependence on others and independence from others. It essentially promoted interdependence of tasks in formal organisations and interdependence of relationships in social, family and community settings.

Exercising authority became a cooperative exercise in achieving goals. In the technological ethos more than one authority could coexist in the same system. The process of exercising authority became participative. The nature of authority shifted from parental dynamics to the dynamics of peers and colleagues. Instead of injunctions, as in the case of family, agreements emerged as the basis of action. The technological ethos generated the operation of authority through shared power rather than concentrated power anchored in one individual. It demanded from individuals that they remain alert, constantly scanning and appraising the system in which they worked. Individuals had to take the burden of assertion rather than practise self-negation; they were called upon to make their role and not merely take a role. Rather than choosing from the given alternatives, they had to seek new ones and bear the consequences of their choices.

With the spread of education and regulated age of entry for education, a peer group and peer culture began to take shape. In a culture and society where age, experience and hierarchy had played a significant role in relationships, the impact of peer group and peer culture began to be felt. For the first time the concepts of voluntary membership and choice of systems to which one could belong emerged. The current generation began to sense the assumptions of voluntary membership in systems. They recognised that inherent in the concept of voluntary membership were processes promoting active participation in systems to influence goals, direction and growth of the system. One of the significant constituents of voluntary membership was the ability to accept invalidation of one's stand and the courage to review and modify it. This shift in identity from social meaning to work required new definitions of interpersonal relationships and authority. It demanded commitment to review, renew and appraise current reality constantly. Without these assumptions and processes, situations were rendered into symbolic analogues of the past and individuals were reduced to objects. New processes were necessary to stop the wholesale transference of familial models of primary systems to people and situations in the secondary systems.

Nature of Work

With industrialisation, the concept of secondary systems became a reality. Here work, linked as it was to education, acquired a different shape. In the agrarian society each individual worked alone and in isolation, whereas work in a formal secondary system was no longer carried out in isolation. Individuals allotted tasks were required to work in a group. Work acquired the concept of teamwork. Joint responsibility became essential for the successful achievement of tasks involving technological complexity. The role-bounded pattern of doing the job, typical of the agrarian society, would prove disastrous for a technological system. Individuals in the new ethos need to cooperate with others and extend themselves. Inherent in the new ethos are three kinds of responsibilities in an organisation. Besides the responsibility for a particular job, individuals share in the lateral responsibility for coordinating their job with other jobs regardless of the levels of hierarchy in the organisation. They also have a share and a role to play in the area of corporate responsibility. This shift requires a change in role taking. Individuals take active roles in the system and are no longer only performers.

The technological ethos by its very nature requires a complex integration of tasks involving different kinds of technological expertise, to be performed simultaneously or sequentially. The system defines several sets of work. Each set requires for its completion different technologies and abilities from a particular group of people. This makes a new kind of demand on the individual, that of discovering his relatedness to the task rather than his ability to carry out specific jobs or work activities. The myth of alienation created by the Marxist nostalgia for a producer society has not only pinpointed this aspect of the technological ethos but has also labelled it evil. The Marxist thesis has mistaken the two thousand years of agrarian patterning of man's life as reflecting his basic nature which in turn has been responsible for the pathos of alienation in society today. In its attempt to decode the concept of man and collectivity inherent in the technological ethos, the Marxist thesis erroneously treats technology as an instrument and a means and then evaluates it from the perspective of the agrarian ethos. Nor have the Marxists looked at the existential quality of man's being. They have taken his conditioning by the agrarian ethos to be his existential quality. For it is true that unless the core assumptions of the technological ethos and the nature of authority, interpersonal relations, and work are made congruent with the existential quality, the pathos of alienation will continue to persist.

The use of technology is an evolving process. The changes in technology tend to shuffle and reshuffle the work design, work flow and task system. In a technological ethos both the individual and the system have to accept the fact that the quality and interlinkages of work will continue to change. This makes all individual work roles or organisational structures invalid over a period of time. The same is true of any work system in its techno-social aspect. The individual and the system both have to add to themselves constantly, not only the skills and techniques but also relevant attitudes and processes. It is only through this process of renewal can the work system remain healthy and dynamic. This means that there is a constant need to redefine the identity of the organisation. In addition to accepting the burden of self-invalidation from time to time, the individual has to learn to make himself dispensible to each role he occupies. Self-renewal thus, emerges as one of the most important assumptions of the technological ethos.

Institution of the Family in the Technological Ethos

In the agrarian society the family tended to be multipurpose and a comprehensive group for the individual's existence. It was the locus of all essential activities. The individual was born there, grew up within its confines, did his work in its fold and spent most of his recreation period in the family and in the community. With the advent of the technological era, the meaning of home changed for the current generation. It was no more a comprehensive, multipurpose group. It had become just one of the many significant locales in his life.

Physical proximity beyond early childhood was no longer essential. As the significance of education increased, the child was sent to school from an increasingly earlier age. With both parents increasingly working, the concept of creche care evolved. This meant the child was sent away at a very early age. The experience of family became timebound. Children learned to be part of multiple secondary systems at an early age. To the current generation the continuation of family closeness beyond a point was experienced as suffocating and oppressive. It became a barrier to the development of the individual's identity. It seemed that with this shift and resultant transition, it was essential that families re-examine their roles and the significance of parenthood vis-a-vis their children. The creation of psychological closeness was seen as the new task of the family. Physical separation became not only acceptable, but desirable and sought after. Parents were expected to recognise the growth of the child and accept him as a responsible being. He could then depart with trust from the family by dissolving the son's or daughter's role, hoping for a reintegration as a person. The family in a technological society has to be conceived as a psychological system rather than a well-bounded social system.

The core of many intra and interpersonal conflicts of the new generation lay in relation to experiences with the family. Most of them felt overcontrolled and possessed. They felt cajoled into remaining dependent. They were denied the opportunity to test reality and discover their own strengths and limitations. They also felt that the family did not accept their maturation and doubted their capabilities and sense of responsibility. The individuals often resented and sometimes felt strongly ambivalent towards the stifling love of the parents. They felt guilty for resenting the role because it meant rejecting nurturance and hence parental love.

The current generation through its struggle with the family had become aware of significant differences between a psychological and social system. The social system, by its very location in the agrarian society, implied location of authority external to the self and denial of representativeness in the system. It had to be essentially normative and needed to be regulated by roles to which people surrendered their authority. The individuals of the third wave of the current generation came to sense vaguely the model of the psychological system needed for preserving a personal identity, equality of the self with others, and the freedom to make new role patterns rather than remain in socially or otherwise defined patterns of roles.

Socialisation of Emotions

By this time three generations had grown up in free India. These decades had also witnessed the stabilisation and consolidation of the technological ethos. Families had acquired a nuclear status. Joint families were beginning to acquire a quality of 'once upon a time' or history. Many participants in the groups came from nuclear families. They lived in neighbourhoods which had no ties with the extended family or the kinship system. The setting of their growth was a close, nuclear family in the midst of neighbours who were strangers. The loss of the extended family deprived them of the network of role relationships and opportunities to distribute their emotions among many roles and people. They also lost the opportunity to turn to surrogate adults in moments of frustration and rejection from the primary object of love.

The small nuclear family frequently lives in a neighbourhood which is not a community but a set of isolated households. Consequently, intense emotions have to be discharged within a limited number of family roles. The implications are that each family member has to learn to manage several and perhaps contradictory roles within the family. For example, in the absence of the father most of the time, the mother had to take on his role towards the children. This increased the problem of double bind on the one hand and the intensities of interpersonal relationships on the other. It made the relationship very fragile.

To help the growing child to socialise his emotions, it was necessary to redesign the family as a psychological system. In the agrarian ethos of role absolutism, with the family acting as a social system, allowance was made for the distribution of raw emotions. Availability of surrogates allowed for their expression. Hence conditioning of individuals to role appropriate feelings was possible. The emergence of the nuclear family created serious problems. What the individuals expected from the family was also psychological relatedness, where each individual could play multiple surrogate roles as well. They expected a climate where they could learn to express openly their feelings and actions without fear of punishment and rejection. They also expected the family to generate processes to help integration of the self and the role. The family was seen as a testing ground on how to make a role. This unfortunately was not available in most families.

According to most individuals in the groups, the significant aspect of socialisation of emotions revolved around the ability to take 'no' for an answer or say 'no' to a request. A confrontation with 'no' in relationships created a psychological threat of having no belonging system which rendered most individuals incapable of saying 'no'. Attempts to say 'no' to a request or a demand aroused feelings of anxiety, fear of rejection and created scenes of temper tantrums. An individual who received a 'no' often had the childish desire to harm others; this frequently generated further anxiety and fear. The current generation attributed this universe of feelings in encountering 'no' in relationships to the quality of family processes, which denied them the opportunity to confront their own emotions and come to terms with reality. In keeping with the technological ethos, the family was conceived as an institution where the individual could learn to take the responsibility for relatedness with the world. According to the members of the new generation, the family fostered the experiences of protection, while denying them the space to express and manage their raw emotions. We suggest that the processes of transition have introduced changes in the meaning of parenting. Most parents themselves were often unsure as to how to behave and relate to their children. A whole range of parenting literature from the West defined new norms and modes of relating. Popular psychology on the consequences of parenting began to be available. In essence, parents were squarely caught in the middle between the commonsense approach of the tradition and the scientific norms put forth by a whole range of parenting literature. This tentativeness and uncertainty seriously affected the children, especially their sense and meaning of belonging to the family.

Change in the Meaning and Quality of Relationships

The technological ethos brought assumptions of a new kind of social design along with a shift in the meaning of work and locus of identity. The assumptions as discussed above, created a thrust to initiate action in the self rather than in the role. The implications of this shift were very vague to begin with. The nature of the shift became clearer as individuals began talking about exploring the contours of interpersonal relations and issues experienced by them. The agrarian ethos had allowed only minimal scope for generating voluntary relationships. It only fostered given relationships. The great mobility in the technological society led to an increase in the number of nuclear families. Individuals began entering and living in communities of strangers coming from different cultural and ethnic backgrounds. The dominant quality of this world was heterogeneity. Consequently, individuals in the nuclear families had the option to either build voluntary relationships or live as isolates. Mobility also meant leaving a set of relationships every few years and starting all over again in a new place.

The concept of collectivity in the technological ethos changed from a set of people who were dominantly permanent and stable in location and consistent and continuous in their lifestyle and modes of living, to a set of people who were mobile and not consistent and continuous in their lifestyle and modes of living. This came with rising incomes or rising status in the socio-economic hierarchy. The centrality of transactions shifted from the primary social group to the secondary social group. The technological ethos did not allow scope for unilinear roles and demanded that an individual play many roles simultaneously. Similarly, increased mobility allowed no scope for an enduring relationship with a fixed community and collectivity.

Most individuals in our groups attempted to build voluntary relationships but found themselves vulnerable. They faced serious problems in experiencing and maintaining intimacy. In their attempts to experience intimacy they confronted terrible doubts and acute distrusts. At the back of their minds they anticipated rejection and the possibility of being exploited and/or used. These feelings rebounded to arouse feelings of lack of self-worth and inadequacy. They were afraid of gossip, betrayal and ridicule. So they chose to relate only superficially. They carried the burden of loneliness behind the stance of 'hail fellow well-met'.

Within themselves, the individuals experienced a strong desire for closeness and intimacy where they would feel wanted and significant. They sought permanence through possessiveness and exclusiveness and at the same time were afraid of being possessed. They came in contact with different sets of people and ended up by having friends to go to movies with, friends to play bridge with, friends to 'bull' with, friends to do tasks with, and friends to hang around with. They did not have intimate friends nor a stable space to be in. They were like wanderers. If by chance intimacy did develop, they lived through the whole drama of adolescent, raw emotionality. According to their account this involved 'testing each other; demanding inclusion in every event and sphere of relationship; a comprehensiveness of relationship which was characterised with the need to confess and know everything about each other.' This eventually led to breakdowns in relationships. The whole system of interpersonal relations became a dream. As we processed these intense feelings in the group, we began to visualise and speculate upon the nature of relationships that could be satisfying in the society of technological ethos. The nature of collectivity implied in the technological ethos created the setting where relationships could exist in task situations. The situations could be social and the tasks could be community or organisation tasks. Often the word *matlabi* (quality of relating only when there is a need to get something done) was applied in a derogatory sense to describe these who acted in the new mode of relationships. The orientation of agrarian ethos, by labelling new modes of relationship as *matlabi*, suppressed the qualities of relationships coded in the technological ethos. The emergent data suggested the following assumptions and qualities of these relationships.

Each transaction in the relationship was complete in itself. It did not matter whether one met the same person again or not. What mattered was that each transaction should be a deep and satisfying relationship and its quality could not be possessed and held in a fixed mode. In each transaction one had to discover a stranger in the other as well as in the self to relate to and with. The joy of a relationship lay in the discovery of this stranger and the ability to relate with new meanings with the same person. All relationships, as such, had to be historical in nature. Expectations from one transaction could not be carried over to the next transaction. One had to learn to walk away from each transaction without a sense of loss. Future transactions could not be mortgaged to past experiences and expectations. The purpose of relatedness had to be held clean in the mind and in the feelings. This meant that each individual was more than his action and manifest behaviour. Individuals were autonomous beings and had the freedom to state their feelings and expectations, but could not impose on other relationships. Essentially, an ability to take 'no' for an answer and say 'no' was essential for the sustenance of relatedness. Thus all relationships had to be open-ended.

The pull of these assumptions of interpersonal relatedness were experienced by the individuals in the groups. It was reflected in their occasionally fragmented behaviour. They could not consolidate these assumptions in systemic or consistent patterns of behaviour but experienced turmoil at the intrapersonal level.

The translation of these assumptions into boundary conditions demanded that the individual internalised not only a sense of security and self-worth but also an ability to distinguish between action and person. Action, after all, was a choice where an individual attempted his best to achieve synchronisation of feelings and purpose on the one hand, and a correspondence with the situation and other individuals on the other. In the process of transitioning from the agrarian ethos of social living to an emergent technological ethos, some time is required for changes in assumptions to take place. In the initial phases unfortunately most of us carry a cognitive map of situations and people from the past. At the slightest stimulus the past maps are revived, masking the quality of the emergent relationships in the context of the current situation. The tendency to analyse actions in terms of motivations and intentions and to locate the source of action in the individual not only put individuals on the defensive, but further blocked them from reality appraising.

The current generation, though implicitly responding to the push of the technological ethos in this sphere of life, found it difficult to accept the quality of viable interpersonal relationships. The meaning of relationship had been coded in acceptance, affirmation, confirmation, and a feeling of being owned by others. The relationship had to be a stable source of fulfilment of expectations for which personal and frequent contact was a necessary condition. The new conception of relatedness—the open-ended relationships—left much scope for the unexpected. Establishing intimacy meant facing the choice between permanence and security on the one hand and the need to trust the individual and give the self and the other freedom to say and receive a 'no' on the other. The current generation sensed the new constituents but found it difficult to cope with the emotional maturity required of them. Neither the family nor any secondary systems provided infrastructures for such a response.

Renewal of Self and System

The technological ethos, demanding a shift in the meaning of self, creating a cumulative rate of change and generating a mobile collectivity, confronted individuals and the systems of the agrarian ethos with their own reality of obsolescence of machines, techniques, skills and roles. It demanded the addition of newer roles and renovation of older ones. In the process of implementation of the technological thrust, those in Indian society came face to face with the drastic discontinuities within themselves and in the system. The new ethos demanded investment in the self and the system to update not only skills but also values and beliefs. For the first time the need arose to internalise the process of continuous self-renewal. The current generation expected their seniors as well as the parental generation to flow with the times and change in a reasonable manner. They expected the parental generation to understand them and their aspirations. It was, therefore, tragic to see to what extent they had been victims of the past, holding on to past modes of relationships. They found it difficult to surrender their expectations and grow mature. Inherent in the technological ethos were processes of reintegration of the self and role and the process of self-renewal. These processes demanded that individuals overhaul their network of perspectives, attitudes and values, and for the first time confronted them with the spectre of invalidation. They never admitted it but the fear of self-invalidation haunted them. They were frightened of disclosing themselves and letting anybody know them. As a protective device they even developed a mark of remaining a stranger. They cultivated superficial relationships with a strange kind of apparent intimacy in strange places. Yet, they regretted this mode of life.

The technological ethos demands integrated persons who can play many roles without losing the sense of personhood. To achieve this individuals of the new ethos had to be sensitive, empathic, self-sufficient and willing to be interdependent. They had to be authentic, responsive and proactive. They needed to be autonomous and capable of interlinking with other autonomous beings without being possessive. It required the ability to relate in a number of ways at many levels with the same individual without stress. Individuals could be and become without negating the being and becoming of others. It required the capability to respond and to adapt themselves to emerging realities, not only in actions but in attitudes and perspectives. The technological ethos demands institutionalisation of infrastructures for reviewing the self and system in order to achieve renewal of both. Renewal processes were the core infrastructures for coping with the rate of change in the technology of production and social living. In the process of transition, roles, beliefs, orientations and perspectives and the link between values and actions become obsolescent. As such, relevant and congruent processes of the technological ethos became essential.

The current generation witnessed many shifts in their life space. Physical and social mobility increased. The role structure broke down. New goal sets were created. A set of voluntary relations were entered into. The internship period was extended and the nature of work changed. Individuals directly experienced this shift and also felt the pull of the technological ethos. The overlapping and lack of clarity of the constituents of this emerging ethos made the current generation anxious, confused and unstable. Caught between the two kinds of ethos, they had to walk a tightrope and often lost their balance and fell. However, some were resilient enough to get up and continue their search, being hurt in the process.

The poignancy of their struggle and the resultant impasse were reflected in the current generation's experiences of these two competing ethos and cultures.

One is the ethos of faith, following the path, living with doubt, living for others, living by roles and of doing one's duty. This is the culture of the agrarian ethos of India. It implies stability, permanence, continuity and consistency. The individuals today live and suffer and yet find it difficult to move away from this ethos.

The other is the ethos of reason, searching, selecting and creating the path, of resolving doubts, of living with others, of living by integration of the self and of accepting responsibility. It is the ethos of change, of uncertainty and of constant unfolding. This is the culture of the technological ethos of tomorrow which the individuals sense and respond to with hesitation and conflict.

One is the ethos of closeness, of dependency, of subordination, of limitations, of self-negation, of replication and of involuntary existence. It is the ethos of the burden of identity. Such is the culture carried by the individuals in their subconscious.

The other is the ethos of loose links, of interdependency, of aspirations, of self-assertion, of creativity and of voluntary existence. It is the ethos of separations and discontinuities, of uncertainty, ambiguity, open structures and open-ended relationships. It is the ethos of becoming. Such is the culture of the technological ethos of the future looming large on the horizon, beckoning the current generation.

For the current generation the clash of these two ethos has created a culture of transience. It was difficult enough to make a choice between the two and for the current generation it was an impossible situation. Most of them tried to hold onto both. In the process some felt torn apart, while others were paralysed. The choice of the agrarian ethos made them feel left behind and the choice of the technological ethos made them feel cut off from their moorings.

It was impossible to choose; both the cultures were self-contained. One was anchored in roles and the other in the self. Focussing on the role left individuals isolated and incapable of action at the level of the self. Focussing on the self left individuals unsupported, without expectations from others and without a sense of validation. As a result, they faced the classical unresolved duality of the role and the self. This duality had many levels as stated earlier. So the choice gave no comfort. Meanwhile, in the state of transience more and more individuals were turning into Arjunas, questioning their commitments, with Bhimas coming alive only in reactions and Duryodhanas indulging in self-aggrandisement.

The choice of the agrarian ethos demanded replication and remaining in the status of being acted upon. It implied the need to retain the same role and identity throughout the lifespan playing a psychodrama whose script was written by others. The contents of the role might change but the processes were eternally the same. If one got tired and bored wearing the same face throughout life, the only option was to move away from the role into a search for the unknown.

In choosing the technological ethos, individuals lived with the constant burden of being better and different from before. Constant recreation of the self and outside interaction sets could be frightening. Individuals could lose the sense of the self. If individuals saw a new face every time they looked in the mirror, they might not identify themselves. They might begin to doubt the reality of the self and marshal all their energies to stabilise the social reality outside themselves.

When individuals and groups were confronted to seek a way out and make a choice, some of them responded, 'Why choose? Let us synthesise.' The plea for synthesis began with Nehru and the initiation of developmental activities. Take the best of both worlds. The protagonist of synthesis could be a socio-economic planner, socio-political expounder, an educationist, an industrialist, or a man in the street. What was frightening was that none of these people were willing to look at the reality of the morbid symptoms Indian society kept presenting in the name of synthesis. The evidence is available in many forms.

Let us state our own bias. Fourteen years of work with each successive current generation, an extensive and an intensive look at their intrapersonal and interpersonal existential world, and a scrutiny of the structural processes of the society have made us weary and frightened of the slogan of synthesis. Every cry to create a little Japan, a little China, a little Germany, a little Russia or a little America reflects the modern credo of Indian society. Its assumptions are—adopt the form, sidestep the relevant processes, and continue merrily with outdated processes of an agrarian society. This could be the cry of a very anxious and insecure person whose only mode of operation is manipulation. Or it could be the cry of a hypochondriac who in order to treat his illness takes the homeopathic, ayurvedic, unani and allopathic treatment simultaneously, little realising that he is defeating the very purpose of the treatment and is creating a tremendous stress in his own body. The question one may ask is: Does he really want to be cured?

Another question arose in our minds: Do the 'synthesis' slogan raisers want India to grow? Synthesis is an easy way out. Any new response, choice of a new direction, and change anchored in growth has a price. Synthesis is not a choice. It is an attitude which makes ineffective compromises. It contributes to the spiral of invisible waste and unreal aspirations. It also sets the stage for coercive measures in the implementation of social policy anchored in an alien model.

The protagonists of synthesis often justified their stand by referring to the Indian tradition of assimilation, absorption and transformation of all peoples by weaving them into an organismic social life. They forget that the strength of the Indian tradition was the psycho-cultural ethos of the classical society. The ethos was based on two unique assumptions: (*i*) about man and his nature and (*ii*) the purpose and quality of the collectivity. The significant coordinative principle of the classical ethos of the Indian social design was lost on the designers and planners of new India. Any attempt to decode the constituents of the classical ethos and its design was turned into a sterile, ritualistic exercise. Having been made out to be a mystery, it had been put aside. It had been overmaligned and ridiculed by large sections of the elite. To others, it had become a speculative proposition for an occasional recall of India's glorious past. No other ethos had found currency or gripped the emotive being of the majority of the Indian population to the same extent. For a country, culture and society to progress, how can the cry of synthesis lead to the development of a healthy society? Every country and culture requires a central coordinating ethos whose dynamism can hold the contradictions and traditions together and provide strength for new directions.

Once again we sat down to understand the antecedents and processes of this transformation. The study of literature and review of our experiences with men and women suggested two independent sources which merged together in the late nineteenth century. The first source lies in the history of India. Indian intellectuals became the victims of a closed system of thought. They learnt to talk the language of *shoulds* and lost their flexibility to appraise realistically the emergent social reality and handle the growing dysfunctionalities in a system syntonic mode. They also surrendered their privilege to make proactive innovations in the social design and refused to use their experience based learning and knowledge in living processes. These pressures tarnished the glory that was India.

Socio-psychological Consequences

Perhaps the degeneration began somewhere in the third century, took roots and gripped Indian society by the sixth century. We can only speculate regarding the its beginnings. Available literature suggests that the seeds of transformation lay in the confrontation between the Buddhist philosophy and its ethos and the Upanishadic and Brahmnic ethos of India. In spite of the fact that Buddhism was defeated in the struggle for political supremacy, its highly logical and appealing ethos continued to flourish. This suggests that it was in the final act of assimilation of the two ethos that the degeneration of the earlier dynamic society began. For the first time the basic framework of the ethos and its authority, the Vedas, were challenged. Thus, the Indian intellectual's essential framework of assumptions about human existence and the redesigning purpose to which they continuously referred, was diffused and contaminated. Further, the attempts of Adi Shankaracharya (A.D. 788 to 820) to assimilate the two ethos and translate them into an overdetermined social design were the next major blow to the classical ethos of a socio-cultural India.

Indian society, however, had not lost its total dynamism and capability to assimilate and respond. Sant Sukhobai in the seventh century, Kabir in the sixteenth century, and many others who were representatives of a stream of reality orientation based on experience continued their commentary, spread their message, and enabled people to handle the growing social dysfunctionalities. But the intellectuals, the representatives of logical idealism, became rigid and refused to discuss and negotiate a new design. The sages could only articulate their sensing of the new social conditions and formulate a new interpretation of the ethos. However, in translating this new interpretation of ethos into new boundaries for social design, the role of the thinkers, adept in the processes of scanning, monitoring, reinforcing and building linkages, was essential. This role lost its vitality and the fine balance of the Indian social design collapsed. Individuals were overwhelmed with anxiety and insecurity. There was turmoil. Recourse was taken to role rigidity for handling insecurity. The self and its social concerns were put aside. Spiritual salvation through devotional commitment was still available and served to neutralise the stress generated by the rigidity of the role. The forces of individuation and social redesigning were obliterated.

The primary assumptions of Indian ethos and social design left the burden of managing primary ambivalence on the individual. The social design only provided boundary conditions and infrastructural processes which could help the individual to forge a creative life space in the system and manage ambivalence to fashion a unique psycho-cultural identity. The individual could adapt and move into the model of a completely socialised role like Rama, or an utterly individuated one like Krishna. Once the intellectuals failed in their guidance, these options offered by the social design were also lost. There was a movement toward the model of utter socialisation following Rama. They once again became only *sons* and their life centred around one single role. Gradually more and more individuals responded this way and the whole social system gradually became a role prescriptive society. Rama, in his son's role became the ideal. And Krishna was reduced to the state of a child and an adolescent lover. Krishna, as the transcender of roles and as a representative of individuation in society, was put aside. This process was reinforced through the adoption of 'middle path' (*madhyam marg*) of the Buddhist philosophy as a way of conduct. In its incorporation into Hindu ethos, it lost its real meaning leading to pseudocompromises.

However, the religio-philosophical ethos and some of the processes and forms of social design of the earlier era survive. The coexistence of rich and poor, dirt and cleanliness, and other such instances of social elements also survive. With the polarisation of the action system toward absolutism of the role and the continued articulation of the religio-philosophical ethos without the connecting links, the two realities did not sit well with today's generation. It saw the hypocrisy of it, scornful, it turned its back both on the ethos and society. The current generation, being the child of two cultures, was also part inheritor of the counterpoint ethos and social design of the West. In its contempt it chose the option of becoming globalised. Some who could not dissolve the absolutism of the role society turned inwards. They planned a lifestyle for self-survival and self-aggrandisement at the cost of society and its systems.

Indian society also did not entirely lose its responsiveness and its assimilative processes. The sant traditions of Nanak, Kabir, Ramdas and others survived. In the initial stages Nanak, through his own experiential system, called for the assimilation of the concept of brotherhood and identity in religious belonging rather than in geophysical belonging. He addressed the experiential reality as given by the supernatural. Nanak claimed his reconstruction as arising out of his inner voice. From the seventh to the sixteenth century Indian society responded to the experiential tradition of Islam as reflected in Sufism and made it the anchor of assimilation of Islamic ethos in the Indian tradition.

The Islamic invaders may be perceived by some historians as conquerors. But they came to stay and created a space for themselves. They did not come to exploit India for the benefit of another nation. They became Indians. This posture did not shift the economic and political power to alien control. The processes of assimilation could be mobilised and the early Islamic rulers encouraged this process of assimilation. Akbar even tried to create a new religion to synthesise the ethos of Islam and that of the Upanishads.

However, the second shift, viz., towards the British, caused the primary conditions of transformation. The British tradition represented an amalgam of the Judaic ethos defining the concept of man, the Greco-Roman ethos defining the collectivity, and the Renaissance ethos attempting to hold the two together. In addition, the British as traders, then colonisers, came to conquer and exploit India, as against the Muslims who came to conquer and settle here.

The religio-philosophical ethos of the Judaeo-Christian tradition held by the British assumes only one cycle of life in this temporal world. This tradition holds that man was thrown out of the Garden of Eden for commiting the original sin. The purpose of man's life was then defined as 'to stand witness to his worthiness in the eyes of God to be redeemed and restored in grace.' This concept of man anchored in a finite existence, laden with the responsibility of acquitting himself of a sin in the eyes of a supernatural power beyond the self, is not only different, but an exact counterpoint to the concept of man in the Indian ethos. The Greco-Roman concept of collectivity implied centralisation of power in a few democratically chosen hands and the marriage of political and economic power. During its heyday, the Roman Empire subscribed to the theory of deterrent punishment and operated within a highly legalistic framework. It tended to place power in a very small superstructure constituting a select few, chosen by birth or adoption. In this stage of the collectivity began the basic seeds of the conflict between 'haves' and 'have nots', leading to revolution and counter-revolution, overthrowing one set of elites in power, only to be replaced by another. The processes of society tended to become static because authority was not decentralised sufficiently.

The British holding political and economic power operated in India with a vision of the above. The Indian masses became serfs for the first time. Even the unit collectivities lost their self-governance by consensus. The British used the education system to promote their idea of collectivity, and actively attacked the Indian tradition in their attempts to colonise India. A systematic campaign to distort India's past and tradition was mounted by the writers of the West from the middle of the nineteenth century. The evidence of this distortion has been abundantly documented by various authors, amongst whom Dharam Pal is the most revealing.

In its early encounter with the British Indian society responded by assimilation. Initially it sidestepped the Greco-Roman concept of collectivity and responded to the Renaissance ethos. A phase of an Indian Renaissance began in the 1830s with Raja Ram Mohan Roy and Swami Dayanand. They were still close to the Indian concept of man and collectivity. They saw the ethos of the Renaissance as a means to counteract the Western dysfunctionalities which had cropped up in society through social and historical processes, and they took to it with a vengeance. It was only much later in the century that Indian society responded through adoption rather than assimilation of the Greco-Roman concept of collectivity.

Thus were our explorations into the nature of the social and cultural heritage of India and its transformation into the current society. For the first time the generation of today has a logical and consistent framework of the Indian ethos, which till now had been made out to be vague, abstract, mystifying and otherworldly. It had always been communicated to them in glorified Sanskrit verses with complete disregard for the links of the ethos with current social reality. It had always been stated in terms of idealistic *shoulds* and without communicating the operational links of the ideal to reality. The forms had been described but the processes of the ethos had been left unarticulated. In fact when we looked at our educational scene in India, we found that no perspective from within was ever given. For instance, in dealing with the caste as a system, various explanations from without had been given. We had been told that caste is a product of the Aryan invasion, a political explanation; caste has been explained as a projection of psycho-sexual pathology (around mother and women), a Freudian explanation. It has been explained as a process for maintaining racial purity, a bio-social explanation. But caste was never looked upon as a possible device for enhancing the dignity and freedom of the individual, or as a means for self-realisation.

As we looked into these aspects of Indian reality, we were struck by the many of distortions of the Indian ethos by foreign authors. We wondered why. It reinforced our belief that education based on the 'English' pattern created distortions. The books recommended to today's generation dealing with human civilisation, such as the history of ideas, of architecture, of physics, of mathematics and others, all trace the development of the human ethos and knowledge from the Greeks to Jews to Persians and then move over to China. India and its contributions are reduced to a passing mention, making India an outright borrower.

We have realised that this is the quality of a large number of writings. Sprinkled in between so many of such writings are a few positive accounts and, as indicated earlier, they go to the other extreme. They glorify India. It seemed to us that a realistic appraisal of India's past, out of which the Indian identity arose, has been missing in the educational system. We found ourselves speculating as to what extent the Indian intelligentsia, in their need for acceptance by the West, has reinforced the stance taken by Western authors.

The quest with which we began to relive the Indian reality—the quest for a perspective on values to live by, for a perspective to come to terms with existential themes of identity—needed a realistic appraisal of the past. We wondered why such an effort had not been forthcoming. In our meetings with the new generation we discovered that the impact of this stance of the intelligentsia was that Indians carried the burden of embarassment, self-deprecation and self-hate.

No wonder the new generation is willing to disown its own culture and cut itself adrift only to attach itself to the so-called *foreign* culture. How tragic is the lot of this generation? Can they ever feel alive? Can they ever develop commitment? And what does the future hold for a culture and a nation whose citizens feel defensive and apologetic; or at best fiercely proud about their past heritage, their origins and their beginnings, all the while trying to hide the hate, deprecation and embarrassment regarding the present? We see no future for a culture, a society or a nation or for those aspiring for a little of Japan, a little of China, a little of Germany, a little of Russia, and a little of America—a little of everybody's but its own. All these countries are what they are because individuals of these nations have a wholesome and a realistic respect for their past and their traditions. They contribute to its persistence and continuity. The dynamism of their cultural identity is squarely anchored in the positive but realistic appraisal of their own cultural past. We know of no culture in the world that has come to the forefront and which has achieved success without acquiring a wholesome respect for its heritage. If we take into account the response of the current generation, it seems that the Indian elite is not only blind, but discourages the realistic appraisal of its cultural heritage.

What did we learn from attempting to relive the distant past of the religio-philosophic ethos and social design of India? It is difficult to give a complete answer. But one thing was clear that the reliving toned down our resentment and that of the participants in such an exploration and removed most of the reactive feeling and induced hate. Besides, it helped prepare the participants in the groups to seriously explore the past hundred years of Indian society. This gain in itself was tremendous. Earlier any discussion of the current Indian society triggered off only a deaf man's dialogue, but now the quality of explorations had changed and experiences flowed freely. Earlier attempts to generate a dialogue made individuals fly off at tangents. They talked from their level of preoccupation which was, as we discovered later, nothing but their particular existential theme and identity dilemma. The reliving of the past stopped the *deaf man's dialogue* and made it possible to explore the identity themes.

It became clear that the constituents of the agrarian and technological ethos were unique. These unique elements were reflected in the family, educational, work and other institutions and traditions of society. It created new structures, forms, role taking processes, and modes of relating with people, both in the family as well as others. At the societal level it created agrarian–technological and rural–urban distinctions. The collectivity of Indian society did attempt a synthesis. Let us see what the constituents of this synthesis were and what emerged from it at the level of the individual, collectivity and system. Chapter Six deals with all of these.

Six

The Nature of Indian Anachronism

Indian society today is confronted with two dominant types of ethos which are contrary to each other. The country has invested enormous resources and societal energy in trying to synthesise the two. However, attempts at synthesis have unleashed processes resulting in many dysfunctionalities. For a deeper understanding and discovery of the new responses we need to look at the cultural metaphor of India and its underlying processes. These metaphors and processes helped create a society of diverse communities held together in a macro-cultural framework. Attempts at synthesis have eroded this macro-framework in which the diversities cohered, resulting in two kinds of fragmentations, one societal, the other intrapersonal.

The intrapersonal fragmentation has created a set of modernised elites, who think and act according to Western metaphors. They also hold continuities and consistencies from traditional Indian society in their structures of feeling and interpersonal relations. There is a second group which still adheres to modes of thought, action and interpersonal relations which are traditionally Indian.

The societal fragmentation created by attempts at synthesis has aroused a community specific ethos and the search for a distinct competing identity, thus leading to the eroding of the cultural framework. It has weakened the integrative processes of our society. Evidence of this is manifest in the current Indian scenario. There are increasing demands for dividing the country along parochial, linguistic and other kinds of boundaries, creating social distance and tension between people from diverse communities.

A look at Indian cultural history suggests that the metaphor of synthesis is neither symbiotic nor syntonic with the basic metaphor of Indian culture—creation of a society of diverse communities in a unified framework. What is this cultural metaphor? Before we examine this, let us look at the metaphor of synthesis itself. This metaphor is borrowed from the Hegelian dialectic and its basic assumption suggests that the process of actualisation of any 'thesis' in society generates several forces which eventually acquire dominance and lead to the crystallisation of an 'antithesis'. Conflict between the two then generates processes for a new synthesis. This synthesis becomes the new operative thesis which leads to the dynamics of generating yet another antithesis. The process continues. This is one view of progress in society.

We asked many people their understanding of synthesis. We reproduce below the three ways of understanding synthesis as held by a variety of elites and common people.

One: Take the best and well-tested idea, method, technology from anywhere in the world and put them together to create a new dynamism for our society.

This concept of synthesis has made us borrowers from all over the world. What we have borrowed are only the forms. The contextual and infrastructural processes that generated the idea, technology or method have been ignored. People who subscribe to this view seem to assume that ideas, technologies or methods have no specific roots in the cultural contexts and infrastructures of their parent societies. As such, these borrowings have created superstructures without solid foundations. Enormous resources and energy are then invested to sustain these superstructures. This leads to a lot of invisible waste and creates unintended consequences which are very often pathologies or dysfunctionalities in individuals and society.

Two: Let us preserve and bring forth the best of our culture. Let us also continue to borrow the best from outside and the two will eventually merge into a new whole.

The proponents of this view are willing to ride two horses, with specific consequences. On the one hand, the view tends to construct regional or ethnic cultural identities out of the collage of diverse cultures inherent in Indian society. This leads to fragmentation. Each Indian community has tried to claim the purity and sanctity of its own culture and, at the same time, seek resources to grow toward modernised borrowed forms. Indian society today is embroiled in the struggles of fragmentation around cultural identities and competition for resources. This has led to turbulent political processes which have endangered the very nationhood of our society.

Three: Let us use the input–output model of the Western legislative and socio-economic programmes based on rational designs to bring about relevant changes in the society.

The proponents of this view believe that reality is defined only by three coordinates. These are technoeconomic, scientific and socio-economic. They do not consider it important to look at the psycho-cultural and psycho-spiritual aspects of the society. Seventy years of Soviet and forty years of Indian socio-political experimentation have shown that these aspects of a society can neither be wiped out nor ignored in development considerations.

This view is anchored in nine basic dimensions of the dominant Western ethos that has been cognitively internalised by the educated elites of India. We will return to the exploration of these nine dimensions a little later.

We reviewed our understanding of the meanings of the construct *synthesis* with the participants. It suggests that synthesis is not the answer for our society. It was interesting that some participants from time to time provided examples of their understanding of synthesis. One of them narrated a story. He said that in a temple in Orissa, there is a wall painting of a woman. The woman in the painting looks grotesque and horrible as a whole. A close examination, however, reveals that the artist has painted each part of the woman in terms of the poetic analogy of beauty of each part. For example, a parrot-like nose is considered beautiful. So the nose of the woman in the painting is actually the beak of the parrot. Fish-shaped eyes are considered beautiful. In fact one of the most common names, Meenakshi, is derived from this analogy. So the eyes of woman are shaped like a fish. The ears, eyebrows, the neck, thighs, waist and breasts are painted to depict ideal poetic similes. The total integration is, of course, quite ugly. This is how the participant experienced the first type of synthesis. Similar experiential examples were provided by the participants for other two types. They are too unpleasant to describe; suffice to say both have very strong anal imagery.

Together with the participants, we searched through India's cultural history to identify the cultural metaphor of maintaining dynamism in the face of a continuous influx of diverse world-views, religions, lifestyles and technologies. We all agreed that over the centuries India had succeeded in evolving and stabilising a social design which allowed manifest diversity to coexist across communities and yet converge and cohere in a broad cultural framework. Essentially, the Indian social design created a 'culture state' rather than a nation state. The social design has been discussed at length in Chapter Two under the section *'Sthitpragna'*. However, the roots of this social design, its parametric conditions and its operating processes were grounded heavily in certain macro-frames and metaprocesses which link the social structure to the cultural collage and ethos of Indian society. The social design, though bound by certain parametric conditions, was always evolving in response to changes in the interfaces of people, communities, and religions with their environments. In fact, this recalibration was consciously done from time to time. According to Acharya Chatursen, the effort was often led by kings. The *Manu Smruti*, the basic Indian codification of social design, was recalibrated through the organisation of the *dharma yagna or rajsuya yagna*. At the time of this yagna, the king invited not only many other kings to pay homage to him, but also invited sages who were to look into the social dysfunctions that had appeared during the previous twenty to a hundred years. The sages discussed, debated, and finally came up with recommendations which were proclaimed as the new codification. Thus a hundred and eight other *Smrutis* came into existence each of which comprised the material consolidated by the sage entrusted with the task. What is important to remember, however, is that these recalibrations of the social design never challenged either the macro-frames or the metaprocesses of the Indian ethos.

The metaphor was not altogether unknown. Labelled *samanvaya*, it has been mentioned by many philosophers and writers in the twentieth century. It may be analogously translated as 'assimilation'. The *Random House Dictionary* defines the word 'assimilation' as:

> Physiology: To convert food into substances suitable for absorption by the body system.
> Botany: The total process of plant nutrition including absorption of external food and photosynthesis.
> Sociology: The merging of cultural trends, not involving biological amalgamation, from previously distinct cultural groups.

Another variant of the meaning of the word 'assimilation' is: act or process by which a sound (quality) becomes identical with or similar to any neighbouring sounds or qualities in one or more defining characteristics.

According to Sanskrit grammar the word *'samanvaya'* is derived from the words *'anvaya'* and *'sam'*. The word *'anvaya'* has various meanings.

> Grammar: The natural order or connection of words in a sentence.
> Relationships, family, lineage and hereditary. Here the emphasis is on continuity and its unfolding in terms of origins of people from the same roots.
> Logic: Statement of constant and invariable concomitance of middle and major term.

'Sam' means simultaneously, all at once, at the same time, together; a mixture, like butter-milk and water.

Put together, the word *'samanvaya'* is defined as a regular succession of orders or mutuality of unfolding.

The meanings of these words suggest that something external or alien, when introjected by a living organism, undergoes various processes and is converted into forms which are bio-symbiotic and bio-syntonic to the living organism. In the process the foreign element loses its identity to the body system of the living organism or, as in case of a society, within the macro-frame of the cultural identity. That part of the foreign element which is not assimilated is rejected as waste by the body. If retained in the body, it becomes a source of pathology ranging from general disturbance or discomfort to cancer. The body metaphor also involves chewing, ingesting, digesting, and further conversion of the foreign element into acceptable constituents of the living organism. For society and culture, such retention leads to conditions of ennui. *Samanvaya* or assimilation involves a systemic coherence which can selectively process diverse elements of external or alien ethos into elements having resonance with the existing ethos.

The second aspect of the word *'samanvaya'* or assimilation in the sociological sense implies that variants in characteristics and qualities are adapted through establishing resonance and broad matching at the process level, not just at the form level. Having picked up the major metaphor and its meaning, we asked ourselves what were other words connoting the various boundaries and processes that operate with and behind *samanvaya*. To us it appears that the Indian ethos had an existential macro-frame and a set of metaprocesses that determined the nature of *samanvaya.* This existential frame and the metaprocesses had to primarily manage the inevitable interface of the culture and society with the environmental changes over which the ethos had no control.

The macro-frame is grounded in certain basic assumptions.

1. All elements of the universe have a life-force. These elements may be hard-core matter, trees, animals or human beings.
2. The life-force crystallises into multiple forms of living things. This is dependent on the life-force interacting with the ecology.
3. Crystallised life forms can be transformed and transmuted. They can evolve into more distinct forms or groups of forms. In the case of human beings, the transformation and transmutations occur through reincarnation.
4. All living things, diverse in quality and temperament, and all human beings, diverse in their beliefs, lifestyles, modes of thought and action have a right to exist, claim a space, and flourish in their uniqueness. Preservation of each unique tradition and style is part of being human.

As such, the social design evolved in India was able to create a life space for each manifestation of life forms. Within this space while they retained their distinct cultural identity, they were assimilated and integrated in the total structure of the culture state. The societal process of creating the space for each diverse form has been described in the *'Sthitpragna'* section of Chapter Two. Creation of the tripodal structure of society was the central structural innovation of Indian society.

In their derivative forms the above four assumptions have been applied to many sectors of life in Indian society. The major Indian sutra *aham brahmasmi* is the final culmination of the above four sutras. However, a culture or society organised around certain definite and invariable beliefs may be desirable but is inadequate for human beings to unfold their full potential. Such a society or culture tends to dominate by the standardisation of life space. It leaves no space for variants and ambiguities to operate and evolve. In fact, such a society is prescriptive and refuses to keep pace with the unfolding of the existential flow and the changing interface with the environment. It must be recognised that culture can at best be a collage, where a wide range of beliefs can converge and cohere from time to time. For this Indian sages and scholars made use of patterns of thinking in terms of associative logic and symbolic analogues of living processes, rather than the linear and logical cause–effect modes of the modern West.

Till about the end of the sixth century, Indian society managed to retain its coherence, continuing to comprehend its diversities in unity and unity in diversity. Perhaps the erosion of the basic processes of assimilation began with the society's failure to establish a correspondence and coherence between Buddhism as the organised religion and the Vedic ethos, which was reflected in the decentralised groups who have now come to be known as Hindus. In the next six centuries Indian society struggled to create a new ethos through the attempts of a whole set of poets and scholars like Adi Shankaracharya, who to a large extent succeeded in recalibrating the ethos and the social design.

However, in our interpretation of cultural history, it seems that Shankaracharya's attempts converted most of the 'oughts' of the Indian ethos into the 'musts' and 'shoulds' of behaviour. This eventually entrenched Indian society in the behavioural values of living. It undermined the process values of living as promoted by associative logic and pattern thinking. This increased the rift between behaviours, feelings and thoughts. Such a gap is natural in all living traditions where the 'oughts' are stated for the collectivity and the burden of their translation into social action rests with individuals and localised communities. The role of the sages, prophets and messiahs was precisely to help societies to manage the increasing rift between their 'oughts' and 'shoulds'; to engender a dynamic tension for their further evolution rather than freeze them into rigid codifications as was the case in Indian society to a large extent in later centuries. Consequently, they turned to the supernatural and its intervention in daily living. Thus, the sixth century led to the rise of tantrik modalities. To a certain extent this revived many layers of Indian identity including the tribal. Furthermore, this period saw the emergence of many heroes who tried to defend the parochial boundaries of their communities against invasion. From this era date the sagas of Alha Udal, Malkhan, Prithviraj and others.

During this period a new ethos also appeared in India. It first began with the invasion of Sindh by Islamic aggressors. However, they came, looted and went away. The ethos entered more actively around the tenth century, when the Islamic Sufi saints wandered about the countryside. However, Muslim forces began invading India around the late eleventh century and occupied a large part of northern India. By the thirteenth century, Muslim kings were well-established in Delhi.

This meant that from the beginning of the twelfth century to about the end of the seventeenth, Indian society had to cope with the incoming Islamic ethos and political dominance. This confrontation led to a very strange phenomenon. If we separate the two components of Islam, i.e., the Islamic ethos represented by Sufism and Quranic fundamentalism as practised by the maulvis and the political wing, we can see the country caught up in many ambivalent processes—reverence and hate, closeness and distantiation, joy and anger, tolerance and revenge and many other such contraries.

What happened during this confrontation requires very comprehensive documentation, much of which has been done by several scholars very effectively. However, we are only reproducing here what we found alive in the feelings of our participants.

During the next five to six hundred years Indian culture and society were faced with the dynamic but also aggressive stance of Islamic culture and its political wing. It also came into contact with the world-views, beliefs, and practices of the highly ascetic and mystic traditions of Sufism. While the ethos of Sufism had a deep resonance with the Upanishadic culture, the aggressive stance went against the earlier Indian tradition.

The symbiotic and syntonic link between Sufism and Indian tradition set in motion an era of creativity which was profusely evident in the development of many new forms in the field of art, music, literature, architecture as well as many modes of living and social etiquette. It also led to the creation of a new language. At another level these assimilative processes were evident in the attempts to forge new sets of religions. One cannot forget the contribution of Amir Khusro. His illustrious name was followed by Jayasi, Rahim Khankhana and others, who wrote poetry in Brijbhasha about Hindu gods and heroes. One cannot forget Guru Nanak, who attempted to create a new religious ethos by integrating the Sufi tradition with Hindu bhakti. Kabir continued this tradition. Even the great emperor Akbar tried to create Din-Ilahi, a new form of religion. His grandson Dara Shikoh translated the Upanishads into Persian. However, it seems that every attempt at assimilation was countermanded by the rigid and entrenched ritualistic tradition of the maulvis who supported the political domination. Thus one could say that India survived as a political entity with diverse communities, but its cultural structure suffered a great setback.

Even before the Islamic ethos and faith could find a place in the macro-cultural framework, Christianity came to India with the Europeans. The Portugese were the first to enter, then came the French, followed by the British. The British finally established themselves as a political power. Successful in incorporating India into their empire, they managed to create a political nation state. In this way the fabric of the European ethos penetrated India. Here again there was a split between the political dominance and the message of Christianity. The message of Christianity had many resonances and continuities with the Indian tradition. This was evident from the fact that as early as A.D. 52 Indians did turn to Christianity. However, when backed by political power and its misuse, it created an ambivalence similar to the one engendered by the Islamic invasion.

From A.D. 1750 onwards, Indian society experienced major transitions and a very significant sense of transience. It had to wrestle with three dominant types of ethos. Firstly, the Vedic ethos as held by diverse Hindu groups which was no longer convergent and coherent. There were too many differentiations and interpretations for the management of social behaviour. However, these differences did not erode the social design of Indian society in any major way. The second was the ethos of Islam which challenged society by its power. However, in terms of the social design it merged with the Indian ethos and created enduring, though somewhat ambivalent, relatedness. The third was the Judaeo-Christian ethos as fashioned by European societies. Once again, the primitive Christian ethos had very strong resonances with the Indian ethos. In fact some charismatic Indian leaders were greatly influenced by it. The process was very similar to the Islamic invasion. It heralded another very creative era in Indian society. A new language, its literature, culture and worldview were introduced to and incorporated by upper-class society. The very concept of a nation state took root and a new revivalism resulted. From the 1830s individuals like Raja Rammohan Roy and Swami Dayanand began to create traditions assimilating the worldviews inherent in the Christian and to some extent in post-Enlightenment European thought. Hindu communities were greatly influenced and joined their movements to fashion a new and more relevant world-view. The major reform movements were the Brahmo Samaj and Arya Samaj. In fact the emergence of India as a modern society, and its assimilation of Western knowledge, could not have been possible without the evocative work of these movements. The Arya Samaj became a much wider movement than the Brahmo Samaj and founded hundreds of DAV (Dayanand Anglo-vernacular) schools and colleges. Many prominant Indian leaders were educated in these educational institutions. A cultural context was created in which people like Tilak and Gandhi could later sow the seeds of Indian independence.

To a certain extent exposure to Western technology and learning also brought to the foreground Hindu traditionalists. Their contribution was coloured with revivalism, but in their own way and specially in the field of education they supported modernisation. People like Pt. Madan Mohan Malaviya, Maharshi Kurve and others like them cannot be forgotten in this context.

At first the Muslims did not respond to the move for modernisation through Western knowledge and education. However, in the early twentieth century people like Sir Saiyed Ahmed Khan and a few other Muslim leaders organised centres to bring the larger Muslim educational population into modern education. However, according to our understanding, the rift between the Hindu and Muslim traditionalists continued to frustrate the emergence of any real assimilation. The Western ethos was effectively internalised but not really introjected and assimilated. It became a part of the logic and rationalisation of progress. The emotive aspects of the Western ethos began to be internalised only as late as 1965. It has still not been introjected. This is evident in our study of more than three thousand young people from all across the country during the period of 1971–89.[1]

If one looks at the encounter of Indian society with Islam and the Christianity, it is evident that the assimilative processes failed because of the shift of political power to an alien group. However, there have always been dominant aliens who have invaded India and held political sway on the Indian continent. It began with the Greeks and continued with the Huns and Shakas. All of these lost their identity in the Indian social design and ethos. Perhaps the reason for this was is that the Huns and Shakas and such other groups did not have as

[1] The report of this study is available in Garg and Parikh (1976). Part of this book is also based on the data from these young people about the nature of Indian society.

strong and well-defined an ethos as Indian society had. The Mughals and the Europeans came not only with superior armies but also modes of social design which have survived over years. Moreover, these groups came to stay and made India their home. As soon as the Mughuls accepted India as their home and did not hold it as an enslaved nation the processes of assimilation of Indian society took over to create a space for their communities within the diversity. This seems to suggest that Indian assimilative processes tend to weaken in the face of aliens, who are distinguished by their objective of colonising a nation purely for political and commercial reasons.

However, this does not seem to be entirely true. For example, the Parsis, the Jews and the early Christians who entered India between the first century and the seventh century A.D. came with well-defined religions, ethos and social structures of their own. They were assimilated and a space was created for them without violating their primary boundaries. It seems reasonable to believe that the locus of political power in society is the major cause of the failure/success of the assimilative process.

The lack of agreement between the Muslim and Hindu leaders about political boundaries continued. Part of the blame must be attributed to the British rulers, but we believe that much of the tension and subsequent conflict leading to the division of the country is grounded in the deeper layers of ambivalence reflecting the failure of early assimilation. This unrest was further fuelled by the increasing socio-political awareness of plural communities held together in somewhat coherent collage by the Indian tradition. As early as 1929, many of these communities such as the Untouchables and other minority groups started to fight for a separate socio-political identity based on their claim to a distinct psycho-social and psycho-cultural identity. Thus, the fragmentation in the Indian collage and social design was accelerated. The three dominant types of ethos and the underlying ambivalences of Indian society created the dynamics of a power struggle in the later British years. This continues in a more intense manner today, aggravated by many new inputs since independence.

Thus, it seems that this new situation finally succeeded in eroding the dynamism of assimilation. Individuals found themselves confronted with difficult choices while trying to integrate their identity. Three kinds of processes resulted. The first was revivalism and fundamentalism. The second involved disaffiliation from the normative traditions of the past. This was evident in middle-class people who drifted into the lower levels of professional and bureaucratic services. At one level they maintained the deeper emotive layers of Indian *asthas* and *samskaras* but at another they turned to the logical and rational ethos of Europe. The third process led to the creation of a new class of people who are aptly described as 'brown sahibs'. The label included a wide variety of people. Some of them were the earlier professionals like the Prabasi Bengalis. Others were the entrants to the higher echelons of the bureaucracy such as the ICS, others were lawyers like the Jaykars. This group eventually developed a culture which walked a tightrope between modern elitism in thought and functional actions on the one hand, and emotive satisfaction derived from the expressive arts of traditional India such as music, dance, paintings, etc., on the other. The literature of the period 1910–40 in Hindi and imaginative Bengali is full of the experiences of these brown sahibs, giving a fairly deep insight into the dynamics of a culture of transience which appeared after the assimilative processes and their dynamism were eroded.

In the second decade of the twentieth century Gandhi took the lead in revitalising assimilative processes. He tried to weave multiple world-views into a macro-cultural frame that supported a new social design. He launched what he called the *rachnatmak karyakram* (social constructive work). This was his attempt to regenerate the assimilative processes as well as create evocative symbolic actions and ideas in a coherent cultural frame for India's unity and progress. His vision was to combine the culture state concept of Indian design and the nation state of the Western ethos. In this he pleaded for the revival of cottage industries and smaller decentralised mechanical industries rather than building massive industries and centralised production systems.

He promoted a new educational design called 'Basic Education' which was implemented through the efforts of Dr. Aryanayakam and Dr. Zakir Hussain. It was primarily based on the old Indian theory that education should be free of government control and focussed on learning through living and action. A long list of Gandhi's efforts in this direction could be appended here. But the reader is invited to reread and study Gandhi with a fresh look. Thus, Gandhi created new modes for holding the diversity of India into macro-frames of ethos, perspective and action. However, even a charismatic leader like Gandhi finally failed in this area as the struggle of political independence and the installation of an Indian national government wiped out a large part of Gandhi's real work in India. Today it is still there to guide us, but it is recalled only occasionally, as lip service to Gandhi's dream.

In our assessment, processes of adoption, adaptation and attempted synthesis over the last seven to eight hundred years have not take root in the Indian psyche. Instead a set of cognitive and ideational maps held in logical frames have been created. Furthermore, they have promoted many political and social ideologies and approaches for bringing about changes in Indian society. The lack of deep emotive conviction and psychic integration has made these cognitive maps into sources of dissent and forces of divisiveness in the country. These in turn have created a turbulent society characterised by caste, class, linguistic, religious and ethnic conflicts.

This situation has created many undesirable events. The partition of India in 1947 is one such example. Since then, except for the first decade, Indian society has been rocked repeatedly with demands for separate statehoods based on and governed by smaller ethnic linguistic groups on the one hand, and cries for creating multiple nations out of a single nation on the other. All these cries are accompanied by violent riots. It seems that the failure of assimilation has rendered the country open to fissiparous processes undermining the central structure of the nation.

The historian would perhaps say that this is nothing new. In the past also India has never been a single nation except under foreign rule. The unity of India as a nation under the Mauryas and Guptas was still that of a culture state. Political unity was effected through the allegiance of a set of rulers of small states. The nation was held loosely together by the power of the heroic qualities of individuals, the ability to make war, and generally manage border tensions. It was only under British rule that the concept of nation state really evolved.

After 1947 a whole new world-view, along with technology and techniques entered the Indian scene. In addition to the European tradition, the WASP (white Anglo-Saxon protestant) identity of the United States of America has contributed a great deal to growth in the socio-economic field. Along with it, the world-view and socio-political and socio-economic ideology of Marxism, as practised in the USSR, also entered India. They all have influenced the cognitive maps of Indian youths and elites in different ways. Their impact, in combination with the rising aspirations of the people of a country like India, has been very strong, and can be judged by what is happening to Indian institutions as reflected in the study of Indian youth. These alien world-views have overshadowed the psycho-cultural and psycho-spiritual coordinates of Indian reality and restricted the vision most of the elites and bureaucrats to three coordinates, that is the socio-economic, the eco-technical and techno-scientific. The main consequence has been accentuation of the ambivalence towards belonging, and of the perception that everything foreign is better than its Indian counterpart. It has encouraged the slogan: 'Let us borrow the best; let us borrow what has been tested and found successful elsewhere.' So the credo of Indian society today is on borrowing not only resources but many non-syntonic and non-symbiotic world-views, techniques, technologies and modes of social living. It has reinforced the three elitist categories referred to earlier. Today, however, is not the age of the English, the brown sahibs and the Anglophiles.

On the whole the diverse elements constituting Indian society today have fallen into various categories. These have tended to become either orthodox or polarised as modern. Orthodoxy holds together all the cultural beliefs, practices and world-views constituting the traditional ethos and identity of India, whether of tribal, alien (Hun, Shaka, Muslim) or Hindu origin. They do not always agree but in the face of modernism tend to come together by accentuating their fundamentalism. Thus, modernism includes all the West-oriented traditions of brown sahibs, the communists and now the new entrants from America. Indian society today distinctly lacks macro-frames and parameters of social design which can create harmony out of these diverse and competing sets. Consequently, Indians today experience dilemmas and doubts and lack values and commitment to any contextual consideration. The intensity of this conflict is so great as to attract individuals either to fundamentalist extremism or harmful self-centredness and self-indulgence. Both seem to lack direction or a framework to cathcct in order to create an integrated ethos.

The overriding consequences of loss of macro-frame is reflected in the loss of direction as well as limited psychic and sentient synergy for effective growth and development. Today the forces of growth and development are sustained through heavy borrowings of financial and technological resources. Furthermore, no clear-cut direction has evolved. In spite of various Five Year Plans and tremendous growth, no rational understanding about the nature of a future has emerged. There is a general feeling that the nation, in the midst of progress, is floundering. The processes of making choices and commitments have weakened. The collectivity of our society has lost its ability to exercise discretions in growth.

A serious study of twentieth century literature clearly shows how these two mindsets of an Indian scenario have created the conflicts and deep cleavages in the Indian ethos. Short stories, novels, poems and films have played up the issues relating to old and new and traditions and changes. The psychological struggles of men and women, in isolation as well as in groups, have been very effectively painted. After the 1960s films specially started displaying a change from the sentimentalised portrayal of this struggle to polarised anger and action-based heroics of the individual.

Post-independence India has seen growth as well as the effects of transience and transitions of culture. India's achievement in almost all fields of development, i.e., material, educational, technological and agricultural, are perhaps unparalleled in the history of the Third World. We do not include Japan since it was never colonised as was the rest of the Third World by white rulers. But parallel to this growth has been the increase in disruptions, discontinuities, dysfunctionalities and pathologies. Social and psychological processes antithetical to the coherence and integration of society have gained predominance.

Perhaps the most damaging aspect of the culture of transience has been in the field of education. When the Indian education system was based on post-Enlightenment Western thought it provided considerable ethos of some kind. It created a positive attitude towards independence and social resource. However, since the American influence has made itself felt, educational institutions have become restricted to providing socio-technical, informative knowledge. The whole process of acculturation into some kind of ethos has been eroded. The education system today provides no grounding in the contextual realities of society. The inculcation of the Indian ethos has in childhood been replaced by an influx of children's literature and comics from the West, specially America. This has separated the educated elite from acquaintance with Indian myths, sagas and even history. Today's education leaves many a young person bereft of roots.

Despite the above, it was apparent that at a deeper stratum cultural continuity and being in touch with the ethos, pathos, and myths of

India never died out. This is evident in the response evoked, regardless of caste, creed and religion, to certain programmes on Doordarshan. Besides the *Ramayan* and the *Mahabharat*, programmes on Amir Khusro, Mirza Ghalib, Kabir and those such as *Buniyaad* and others have been very popular. A significant role in this field has also been played by the publication of *Amar Chitra Katha*. Before these consciously designed efforts to get in touch with a Hindu constructed identity, our data suggests, young people growing up in the 1960s and the early 1970s were finding themselves cut adrift from an Indian identity.

There are many ways to assess the phenomenology of synthesis between the two ethos. One of the sources of this exploration is the experiential universe of the current generation. We recognised the fact that the phenomenology of synthesis can be constructed through several theoretical perspectives. However, we preferred to restrict ourselves to the living data of people who transact with the phenomenology of synthesis. This experiential data suggests that India has continued to retain many social, individual and organisational processes from its agrarian ethos. It has also adopted many forms of technological ethos from abroad without the accompanying cultural contextual processes. In this process a culture of transience has been set in motion. It has had its positive gains. India has grown in the material sphere. Many new horizons have emerged which have led to the identification of frontier tasks in the society. The cognitive map of the future has crystallised. Indian energy and resources have been mobilised towards the postulated direction and goals. However, along with these gains the culture of transience has also intensified, sometimes beyond manageability. A set of pathological, disruptive and negational processes have also emerged. These processes obstruct the emergence of any new coherence, convergence and collages in the society.

The implications of the synthesis can be seen at many levels. At the social level there has been a rise in senseless violence, self-aggrandisement and search for power at the cost of national tasks. There is an increased lack of commitment to social action. There is an increase in adulteration and corruption, often leading to the loss of human life. And, finally there is the rise of divisive forces in society, reflected in intercaste, interreligious and interstate conflicts and violence.

At the organisation level the attempts at synthesis have meant lack of commitment in work, excessive role-boundedness, rise in personalised modes of relating across functions, roles and tasks, and weak initiative. The failure to evolve corporate processes and organisations are managed either by crisis, fear and/or anxiety. The need for renewing technology, as in the textile industry, has been neglected. Underutilisation or misuse of technological capacity; lack of attention to maintenance; uneven distribution and absolutism of authority and power exist; holding task systems as personal property; and disowning of responsibility for internal process and designing relevant infrastructures are evident. Increasingly organisations and individuals put the blame for failures on workers, others, structures and government policies. And, finally there is a constant harping on the constraints, limitations and impediments to performance.

The phenomenon of synthesis at the social and organisation level has been sustained by converting the government into a symbol of *maibaap*. Nothing can be done unless resources are drawn from government funds. The faith in philanthropy and individual initiative which was dominant in the earlier part of the twentieth century has been largely eroded. Complete dependence on government resources which are borrowed from foreign sources has become an endemic problem of development.

At the individual level the data from the current generation reflects a very mixed universe of values and attitudes. At one level, they seem to be self-centred and consumption-oriented; the behaviour seems to represent a pattern of extracting from the system. They do not themselves feel responsible for replenishing either themselves or the system. When pushed to analyse this, most of them said, 'We pay our taxes, that is our contribution.' In their mind the logic of wider investment in the community was just another traditional sermon. However, it was interesting to note that they did suffer a sense of guilt, anxiety and inadequacy, accompanied by feelings of lack of conviction and trust in themselves. They masked these feelings with a strange display of aggressive confidence, freedom to disengage from systems, and a belief in their ability to create a space for themselves anywhere in the world. However, it was clear that they were seeking a modern lifestyle and a search for fulfilment based upon organisational status. Once in a while, one could get a glimpse of their acute desperation and loneliness. This desperation seemed related to their not really knowing how to live genuinely in intimacy and affection with people around them. They experienced a sense of isolation in achievement and displayed a fear of any kind of deep intimacy.

All social structures have some operative assumptions. Limitations of resources, building up of regulative processes, stratification of people according to accepted criteria and evaluating them in terms of rewards and control of resources are inevitable in all societies. As such, all individuals in a society experience constraints, barriers and obstacles to their aspirations and dreams. They also have strong inhibitions regarding self-action. In every society at any level of convergence and coherence, therefore, infrastructures and institutions evolve to help individuals manage the tensions and anxieties which inevitably result from membership of any kind of social structure. During the culture of transience, these infrastructures and institutions lose their potency to help the individual. They become merely ritualistic forms lacking any cathexis with the existential flow of the individual in society. During the period of cultural transience some infrastructures and institutions do emerge, but they are short-lived and others come up. The process is that of a new wave wiping out the imprint of the older wave. More enduring infrastructures and institutions do not emerge. Should a set of individuals design and promote such institutions which send down roots across the society, the socio-political power elites tend to curb their development. Their emergence is experienced as a threat to the power elite. History is replete with this phenomenon. Socrates in Greece, Christ in Israel, Dayanand in India, Martin Luther King in America and thousands of others have paid the price of their lives at the hands of socio-political elites of societies going through cultures of transience.

During the late 1930s, a whole set of significant generation Indians lived through with this transience. They learnt to balance the dual ethos in a fragmented manner. Jawaharlal Nehru was an arch example. He wanted to promote the modernism of the industrial era on the one hand and traditionalism on the other. He evolved the concept of a mixed economy. He promoted large-scale industries as a core sector, as well as the development of agriculture and cottage industries. During his lifetime, he succeeded in walking the tightrope between two cultures and India grew by leaps and bounds. After him most significant role holders of the socio-economic and socio-political power systems have failed to give meaning and direction to the forces let loose by Nehru's tightrope walking. They have tried to carry on with, albeit in an empty form, many of the infrastructures and institutions that he and his team founded in the first decade of freedom. However many of these have now become political stunts and election slogans.

Nehru's tradition developed two separate strands. The modern administrative and educational elite were fired with the idea of adapting techno-economic, socio-economic, scientific and secular Western forms to Indian society. The traditional elite concentrated on the revival and perpetuation of agrarian institutions. Both traditions gave no time to society to develop an ethos within the psycho-cultural context. They were absorbed in reformist or developmental action. In addition, the modern elite gradually became borrowers of the Western ethos. The borrowed ideologies initially ranged from Fabian socialism to hard-core Marxism. Currently, the American industrial–military nexus is taking hold of the Indian elite. The modern elite dominated the administrative and educational systems. In the name of secularism and rationalism they ignored the living reality of individuals and systems. In terms of individuals they only looked at need fulfilment. Food, clothing, shelter and literacy became their major concerns. They focussed on the scarcity of natural resources and the uneven distribution of opportunities. In their judgement, India required a drive to increase production in all fields of human activity, from agriculture to industry; and with this in mind they set out to achieve these goals. In the process they colluded with politicians who fostered protectionism and played up to the religious sentiments of the people. They invested their energies and national resources in attempts to eradicate the evils of the social structure such as caste, dowry, maltreatment of women, superstition and economic exploitation.

The approach to the eradication of social evils was by legislation on the one hand and introduction of the input–throughput–output model, i.e., pumping of resources and knowledge, on the other. In their zeal they ignored the fact that these evils of Indian society have been there for centuries. In one form or another they have been the evils of all societies. The struggle against those such as caste, dowry and poverty has been waged in India throughout the centuries, from Buddha to Gandhi. Instead of learning from the past that eradication is a self-defeating strategy of social change, they made it their credo.

This approach of the Indian elite is nothing new. It is a continuation of a traditional struggle against the inevitable dysfunctionalities of any social design. The tendency is first to fight the past and then move towards the postulated new ideals. We think that attempts to eradicate or eliminate existing dysfunctionalities per se lead to meaningless waste of social energy. Only a recalibration of design and adding of new elements can improve the situation. We also believe that every choice, besides creating desirable and intended consequences, also creates undesirable and unintended ones. These undesirable consequences can only be contained or managed not eliminated.

The elite often forget some social realities. One of them is that every society is stratified. The very nature of social structure creates hierarchy, ranking and other types of stratification. Everywhere in the world stratification by birth is a reality. In the West stratification has always been based on multiple criteria ranging from ethnic, religious, inherited wealth, or intelligence. In the USSR, which in our construction was a part of the West, stratification is additionally dependent on political membership. In such societies improvement of one's rank has been possible but it has never fully eradicated the layers of operative stratification by birth. It continues to operate in systematic ways in the society. We have often asked the elite why they bear such an intense, vitriolic resentment towards the Indian social design and the nature of its stratification. As against the traditions of all societies, Indian society was tripodic in its structure. The key elements were the caste, *jajmani* and panchayat systems. Together these three structures managed to contain the major dysfunctionalities of Indian society. The *jajmani* and panchayat systems were weakened under the impact of British rule and by our willing adoption of Western socio-economic and political structures. But the caste structure survived because it continued to ensure a minimal psychological security and identity for individuals. It ensured the preservation of a world-view and a lifestyle around the semblance of integration of work, personal and social identities. Furthermore, to a large extent it still continued to ensure work, though at a minimal level of economic returns, to most individuals who voluntarily or involuntarily took to their caste occupation. The elite failed to take into account the positive aspects of the tripodic social design. They are now compounding their original mistake by reviving the panchayat and (modified) *jajmani* systems by legislative enforcement. Their design follows bureaucratic regulations from Western models. We believe that their implementation under law will release some of the residual pathologies which have not been played out so far.

Similarly, there is enough evidence to show that the input–throughput–output economic model as applied to social aspects in India has been non-productive. Unless the nature of the 'black box' which converts the inputs into throughputs is realistically understood and managed, this will occur. Predictions have proved unreliable when linear input–throughput models are used in human fields, including psychology. The black box is as unpredictable as the old man of the sea in the legend of Hercules. We suspect, however, that the modern elite are deeply entrenched in three sets of Westernisation syndromes. The first set, which deals with modes of knowledge, thought and perception, includes: (*i*) Aristotelisation of knowledge, (*ii*) continentalisation of thought and reality appraisal and (*iii*) missionarisation of beliefs about the Third World. The next set, which is concerned with the government operation and bureaucratic action, comprises (*i*) administration by scarcity, (*ii*) planning by 'comparativism' and (*iii*) governance by 'minoritism'. The third set, dealing with the fragmentation of coordinates of society, consists of: (*i*) socio-economic coordinates, (*ii*) techno-economic coordinates and (*iii*) techno-scientific coordinates. All three sets exclude two of the very basic coordinates of social and cultural reality—the psycho-cultural and psycho-philosophical or psycho-spiritual. The entrenchment of the elite in these three Westernization syndromes is so strong as to convince them that they know the nature of the black box of Indian society. They are not willing to look at it and diagnose it afresh from within. They believe that they only need benchmark studies of manifest data to determine the status of the black box without being concerned about its essentials.

The modern elite of the administrative system have put in tremendous efforts to identify, formulate and undertake frontier tasks and to entrust them to new institutions. However, the structures imposed upon these new institutions, except in rare cases, have been derivatives of traditional colonial administrative models. Their design is determined by modes of control, coordination and evaluation from outside the system. It is primarily hierarchical with many levels of management. The design has no concern with the new parameters of tasks and their interface with the environment. The rules and regulations of the imposed structures are irrelevant even for the internal interfaces of tasks of the institution. Some of these organisations have functioned successfully but only while charismatic personalities remained at the apex. As long as such personalities continue to direct, the institutional and structural aspects of the organisation cohere and are dynamic. After the leader's death or retirement the institutional aspects lose their vitality and the structural part becomes mechanical. This is because the identity processes of the leaders themselves become substitutes for the infrastructures and institutional processes necessary for the dynamism of organisations. After their departure, organisations are left in the hands of people who have been merely trained in performing within the structure. The administrative structures become all powerful and the inheritors of the organisation devote themselves to the maintenance of task performance. Organisations become mechanical systems. Over the last thirty years many dynamic institutions have followed this pattern. The evidence is substantially clear.

The administrative elite do not accept this evidence and continue to design and impose derivatives of colonial bureaucratic structures and administrative practices on all institutions dealing with frontier tasks. Obviously, the colonial bureaucratic structures are either sacrosanct or so deeply introjected that the identity processes of the bureaucrats themselves have become mechanical and frozen. It is interesting to watch the games bureaucrats and politicians play together. Whenever one institution fails or becomes stagnant, they let it continue to exist but go ahead and design another. Eventually, multiple institutions coexist in the same task area. This gives rise to problems of jurisdiction, coordination and activity integration. This strategy enhances the power of bureaucrats but wastes national energy and resources.

Furthermore the modern administrative elite seems to value monolithic, mammoth organisations, be they devoted to production or service. It has been found repeatedly that these mammoth organisations are difficult to manage, and behind their apparently successful operation are hidden huge wastes, high costs of operation, corruption and distortion of the very tasks they have been designed for. Yet, the administrative elite keep on designing mammoth organisations. Reassessment of cultural reality and discovering appropriate models from within is anathema to their own cognitive identity. They operate from a philosophy of borrowing and, therefore, undervalue indigenous resources and expertise.

The tendency of the modern administrative elite to redesign the structure of society by the logic of the task they postulate is counterproductive. We suspect that they have only a vague and diffused concept of the society they want to create. They seem to want to create an amalgam by borrowing little bits and pieces from other successful nations. Grounding themselves in the Indian reality appears to be low among their priorities. They could use the dysfunctions of Indian society as a benchmark for their borrowing.

The situation often reminds us of the story mentioned earlier in this chapter of the Orissa temple painting of a woman. It also reminds us of the story about Darwin who, while visiting one of his friends, was presented with an insect by the friend's children. They had glued together parts from many different insects to make it. When asked to identify it, Darwin, with a serious face, pronounced it a 'humbug'.

Thoroughly entrenched in the comparative framework of the West, the administrative elite direct their entire energy to meeting quantified targets. They invest little energy or resources in building psycho-cultural infrastructures. Even the planning of economic infrastructures has often been an afterthought. Economic development has lagged far behind expectations due to lack of infrastructures. Partial success in meeting the targets over the past twenty-five years has reaffirmed the belief that objectives and targets can best be achieved merely by resource expenditures. Building cultural infrastructures is not, according to them, a significant condition for development. Enmeshed in their own intellectual orientation and the introjected Western ethos, the elite have become immune to the failure and invalidation of their model. They keep on demanding more and more powers to control and monitor operations. Confrontation with failure makes them put the blame on the culture, the ethos and the people of India.

Let us now take a closer look at the contents and processes of the educational system. Since India became independent, the education system has largely ignored psycho-cultural and psycho-philosophical perspectives. Prior to independence, a whole set of nationalistic institutions such as Dayanand Anglo Vernacular high schools and colleges and Christian colleges had nurtured the cultural identity by partially bringing into focus the contents and processes of the Indian ethos. Universities like Banaras Hindu University, Aligarh Muslim University and Allahabad University also maintained, through literary and other activities, links with the Indian ethos while exposing their students to the Western ethos and knowledge.

After independence all such inputs have vanished. Not only were the links with the Indian identity and ethos broken, but even the exposure to the Western ethos has been lost. It seems that the commitment to secularism and rationalism has not only converted the education system into a techno-informative model, but has also crushed the spirit of developing a life perspective by studying the history of human thought. Our explorations suggest that something even more serious has taken place. The professional educational institutions seem to generate negativism, hate and scorn towards the Indian ethos and train students to undervalue modern India's strengths and achievements.

More than five thousand participants from the second and third waves of the current generation issuing out of professional institutions are living evidence of this process. We found them sadly lacking in awareness of even factual knowledge about India. They have not been exposed to the Indian ethos and were unaware of Indian history, philosophy and literature. Their knowledge of India was gained exclusively from socio-economic and socio-political reviews based on a Western perspective, in which India was seen to be at a disadvantage.

Technical education, then, has become so completely techno-informative that it has de-emphasised all elements which bring man into touch with existential pathos and human ethos. In spite of loud proclamations, these premier institutions have ceased to educate or acculturate and have merely become instead centres for technical training.

In its current form this part of the system has failed in its objective of providing secondary socialisation to Indian youth. Education is the only way to create infrastructures and institutions for forging a new ethos. It can provide individuals a space to integrate diverse trends of thought to fashion a cultural identity. Participants in our groups reported different experiences of the college milieu. There were built-in discontinuities between the form, content and the process of education. They were at a loss, not knowing what new patterns of interaction for positive growth they could develop. All fronts were closed and there was no opportunity for affirmation.

This discontinuity in the college milieu, specially in their administrative processes has pushed the youth towards scepticism and distrust. They have felt lonely and anxious. The lack of infrastructures and institutions, and prolonged postponement of their maturity, denied them a context in which to put down roots in society. They had to fall back on their own resources. They gradually became alienated from and uncommitted to the self and the system. Obviously, they then needed inputs to convert the reactive alienation. The only alternative promoted by the education system and the wider society was to conform to the structural norms which were at variance with their own sensing of the living processes around them. This was not acceptable. They had no alternative but to seek a space for experimentation and, through the process of individuation, acquire a core of identity which would allow them a feeling of convergence and integrity within their being and becoming. The education system denied this and even discouraged this search for individuation. Society added to their burden by demanding that they be mere spectators until they began to work. They then condemned them for losing their sense of values.

Due to its inability to give up traditional administrative processes, the college education system retreated from its primary role of helping the individual to cultivate relevant values for living. Instead, it concentrated exclusively on being a storehouse of knowledge, which was disseminated through impersonal pedagogy. Thus, the college system did not provide the students with any models for identification or integration. The student role was, thus, narrowly conceived. The education system and society saw the students' role to be good examinees and so to achieve good results. They were loaded with work and restricted from extra-curricular and extramural activities. Achievement in this narrow academic sense was prescribed as the major goal for them. Such an approach left little scope for intellectual growth. Colleges treated them as future citizens of the country who were currently only immature, irresponsible, indisciplined and impressionable beings, hence requiring control. Their families also treated them in the same way. Refusal to understand and appreciate the youths' socio-psychological world is evident in the behaviour of the important role holders in both institutions. And, strangely enough, the role holders in both systems blame the youth for their lack of communication and the generation gap. Both systems fail to create opportunities and spaces where maturity can be mutually attained. A censoring, fault-finding, demanding, and controlling stance has become the systems' preferred choice. The role holders fear is that the absence of such a stance would corrupt youth. This negative concern for the welfare of youth has made these systems lose sight of their own objective—fostering effective human beings. They have restricted their roles to the making of sons, daughters and students who perform, and are secondarily persons. It is tragic to see both parents and educators spend their energies diagnosing and philosophising about the pitfalls of growth without willing to create a healthy social milieu for human growth.

The problems of the parents can be understood. One can even empathise with their difficulties. After all, their roles are restricted. Their vision is only a reflection of the past. Their primary concern is security. But what of the educators? They are supposed to be in the vanguard, tracing a path for future generations. They play extensive roles in the life of youths. Why then do they surrender their responsibility? Is the genesis of their constructed role their lack of professional conviction?

It seems that the educators themselves are more concerned with the absolutism of the social power dynamics. They have built an armour around themselves to deflect probing looks into the working of their own systems. We have heard educators repeatedly diagnose the shortcomings of youths and systems such as examinations. But they have consistently refused to look at the basic processes governing decision-making within the educational system.

It seems obvious that schools and colleges have provided no direct inputs for the development of identity. In fact, the cultural milieu of schools and colleges has generated some primary conflicts. They have only preached high-sounding but hollow ideals at youth. These institutions are holding onto traditions and practices which may have been academically relevant in the past, but are of doubtful validity today. Consequently, most openly seem to display cynicism towards social concerns and commitment.

The education system in India has become a prisoner of history. Like a prisoner, it is isolated. In its isolation it has become concerned solely with self-preservation. It has neither the awareness nor the willingness to recognise new demands that today's youth confront in the process of their growth. On the whole, the educators live in a make-believe world and are not in touch with the social–existential reality. Unless the education system does some soul-searching, the investment in education will not only remain culturally irrelevant, but also become a source of social disintegration. The emphasis on techno-informative education, without institutionalising processes of value generation, heightens the isolation of youths from social reality. It adds to their scepticism and self-centredness. Growing up in such a system and feeling the stress of their existence, they are likely to become more disruptive and self-centered.

Our explorations into the nature of synthesis and its consequences left us feeling very disturbed. We saw youth after youth struggling to find roots for integrating their identity. They were often desperate. We believe that a synthesis which is neither symbiotic nor syntonic to the culture can only create an undifferentiated mass identity, but never an integrated cultural or personal identity.

Message of the Attempted Synthesis

Our explorations led us to believe that the processes and consequences of synthesising the two ethos leave most individuals in a quandary. They are unable to acquire a simultaneous positive focus on self and system. Many individuals become expert performers but remain socially, intellectually and emotionally immature. Most of the statements made by individuals in our groups implied that they experienced a lack of space for themselves, both in the family and in the broader social system. Some communicated themes can be summarised as follows:

No Situation is My Situation

'No situation is my situation.'
'It is always somebody else's situation.'
'It is the situation of the significant others.'
'They have to decide.'
'They have to make the decision for me.'

This attitude made individuals expect injustice, deprivation and denial from the system. Grumbling became the most dominant outlet for their feelings. Alternatively, the path to get what they wanted from the system was through personalisation with significant role holders and manipulating the situation. When both methods failed, individuals learnt to resort to violence. When we challenged them to take some proactive steps, their answer was, 'It is not our role. Let the system begin first. We will wait and see what it brings for creating simultaneity with us.'

Apathy is the Best Defence

'One should not care too much.'
'I am left cold, nothing excites me.'

One of the participants echoed the two statements above by giving an example. He said, 'I felt nothing when I heard the news of my father's death.'

'Do not get too intimate, you could be exploited.'
'Keep your feelings close to your chest and say only what is expected.'
'Trust nobody.'
'Watch for opportunities.'

We explored this universe of apathy and were surprised by the ten tenets for action that the youth seemed to prescribe for themselves. These are:

Success lies in following the beaten path.
Join the power game.
Use the organisation for personal aggrandisement.
Withhold innovations till you are backed by power.
Merit is not easily recognised.
Organisations want results, not quality.
Expediency is the best policy.
Wait for orders and assignments.
Enhance reward for yourself at any cost.
Do not experiment with new action.

In this chapter, we have recorded the experiences of our participants and have tried to infer the quality of synthesis of the two ethos. Our exploration has led us to believe that whether they be labelled orthodox and agrarian or modern and technological, they are two distinct almost exclusive categories. Their distinctiveness and exclusivity does not provide space for individuals to develop a wholesome identity. The individual and the culture are either fragmented, or become diffuse and lose coherence and convergence in directions and choices. Very often the fragmentation operates in such a fashion as to isolate the emotive and action choices. Individuals fail to develop convictions on which to act. Perhaps they do not acquire even the courage to have convictions. They learn to live by opinions or what is expedient.

If the synthesis is meaningless, what are the alternatives? The answer given to us by folk wisdom was to 'transcend'. We asked ourselves what it is that individuals, society and culture have to transcend? When we explored folk wisdom, we were flooded with alternative statements such as: transcend the past; transcend the living reality; transcend your conflict and act, and finally, in terms of

Gita explore the nature of your commitments and act within those commitments to which you are dutybound. We found it difficult to follow any of these alternatives. Our explorations with the participants finally convinced us that we need to understand the fundamental differences of the two ethos in terms of their underlying concepts of man, the collectivity and their relationship. To some extent we have already explored these concepts in Chapter Two of this book. So we started to seek our way out of the conflicting cultures. These explorations are reported in the next chapter.

Seven

Beyond the Two Cultures: Some Alternatives

During our attempts to understand the synthesis in India and for that matter anywhere in the Third World, we came to some basic conclusions.

Cultures, societies, countries and individuals have made attempts at synthesis—of the old and the new, the traditional and the modern and the East and the West. The resultant dynamics is a culture of transience, held by the individual in his identity. During his growth each individual comes to give meaning to the elements of two cultures. The patterns that a particular individual develops may differ from those of other individuals in the same family. The collectivity has no means, except through indoctrination varying from political coerciveness to religious fanaticism, to make the individualised introjects of the two cultures converge and cohere.

As some of the elements of the two cultures are largely unknown, it makes the output of the synthesis unpredictable. No laws governing the mutations of these elements in interaction are known. Broad guidelines presented by philosophers of history like Toynbee, Spangler and Taggart do provide a framework for studying the understanding of the phenomenon of cultural mutations. However, they neither provide insights into managing the mutations, nor identify processes which, if cultivated, can facilitate the emergence of healthier mutations.

Our attempts have been directed at stating some boundary conditions and their processes which may help develop a framework for managing cultural mutations.

From the available data it is difficult to suggest even a rudimentary framework of a social design which may foster convergence in the culture of transience and thus direct the evolution of society towards a new ethos of man. We have, however, drawn some inferences about the nature of structural and infrastructural processes needed to trigger off a movement towards convergence and coherence of the culture of transience and the technological ethos. We also have had a glimpse of a new concept of man and individual processes, which are congruent with the structural and infrastructural processes deemed necessary for convergence.

The Structural Process

Societal structures and systems have been in the past designed for stability, security and certainty. Indian society, especially its agrarian ethos, reflects the same tendency. This was logical as earlier societies rarely experienced major changes. First the industrial revolution and now the technological revolution has accelerated the rate of change in the environment. This rate of change is also cumulative. The evolution of societal structures and systems lag behind. They cannot be designed and implemented at such a rapid pace. The individuals and groups in society who internalise these structures and systems find themselves inadequate to cope with the emerging demands of the changing environment, and find themselves in the critical position of having to make behavioural responses to these demands. They lack the affective and cognitive wherewithal to determine their responses. Individuals cannot avail of the socio-psychological infrastructures of choice and action. The collectivity appears normless. In order to survive, individuals and groups give in to compulsive action and try to back it up in the name of expediency. This reinforces the rupture between their manifest behaviour and the socio-psychological infrastructures of action. A form of culture of transience is created. The individual finds himself beset with anxieties, guilt, ambiguities, insecurities, uncertainties and all other accompanying feelings at the existential and transactional levels. The individual is deemed 'alienated' and the society is labelled as 'orthodox' or 'normless'.

However, what is really alienated is the structure and the system held in tight control by the legal and formal representatives of the collectivity. This alienates man from himself. This process creates the culture of transience. Individuals, without any systemic process, struggle to establish newer modes of meeting life situations, definitions of their relatedness and choices. The structures, institutions, systems of knowledge and valued precedents of actions become defensive to protect their stability. Their modification is like disturbing a dormant volcano. The culture of transience extends indefinitely. As a last resort, revolution or some charismatic leader intervenes.

An arch example of this process is reflected in the administrative manual of India. Most of its primary clauses were formulated between 1850 and 1900. Each primary clause held sway while a large number of subclauses state conditions of reinterpretation which can be resorted to at the discretion of judges or officers in power. The various subclauses make the choice of action a matter of controversy between the implementing agencies and their role holders in the same government. It also allows space for the so-called 'loopholes' in law which are capitalised on by expedient individuals in society. A through overhaul of the basic assumptions, the basic structure and the system, around new parameters of the individual's relatedness with his collectivity under the prevalent social life, is rarely attempted. The tradition of Indian society to conduct *dharma-yagna*—to repostulate the relatedness of man with his collectivity and recodify it into a new structure and system—is an acute need of the times. This alone can help the culture of transience and build a new ethos of man.

Our data suggests that in order to contain the culture of transience the social structure has to be redefined. Similarly, the relatedness of man with the collectivity has to be reconceptualised. Its design must go beyond ensuring stability, security and consistency, it must have anchors to promote change from within. The responsiveness to the environment has to be coded in the social and organisational design rather than in the leadership. The first tenet for integration of transience and human ethos of the technological society can be stated as: The creation of simultaneity between anchors of stability and change—a change evolutionary in nature rather than a mere adaptation to the existing structural stability of the social design. The tenet implies that change and stability in terms of social organismic process are independent variables. They need to be made interdependent by the nature of social design and its processes. In most concepts of social and organisation design, stability is treated as the primary concern and change as a variant with which to cope. The first tenet leads to two other corollaries:

1. Creating simultaneity between socialisation and individuation.
2. Designing institutions to renew the individual role and organisation identity.

Our data suggests that it is not uncommon for Indian organisations to restructure themselves. In fact, in the first forty years the 'restructuring' of organisations became the dominant occupation of behavioural scientists. Most such attempts, however, are based on the same criteria of stability, security and certainty. In a very short time the structure of the organisation either becomes irrelevant or partially ineffective in managing the organisation–environment interface. Each change in the structure has involved only a redistribution of power and realignment of tasks, but a redefinition of the organisational design and identity has rarely been attempted. Role and task contents have changed. Sometimes new processes have been started but they have remained inoperative. These changes have been ordered but their institutionalisation has not been fostered. This has created a phenomenon where the designed structure has remained on paper as a formal proposal, but the real operation of the organisation has been governed by the emergent structure. A similar process at the national level is reflected where a new constitution has been promulgated, but even after forty years the tasks before the nation continue to be governed by emergent structures anchored in power alignments of bureaucrats, politicians and to some extent the business community. The building of infrastructures to foster the constitution has become the material for preaching and moralising. This has ensured a continuation of the process lag and a culture of transience.

The need of the hour therefore is to develop organisation design for stability and change. The question is, can this be done? We believe that their parameters can be identified and a framework for such a development can be formulated.

Family as an Organisation and an Institution

Let us take the family as an example. Traditionally the family has been designed for stability. The roles are compartmentalised across sex and age. Under sex differentiation, the woman is confined to the role of nurturing the system and managing activities inside the house. Man's role is that of controller and generator of economic resources. Under age differentiation the progenitor has to perform the role of controller of resources and decision-maker. The descendent has to perform the role of an apprentice and a resource generator. The family structure has thus encouraged unilinear role relatedness. Each role has been allocated certain activities of day-to-day living. These are invariable.

This structure is quite stable and functionally very effective in an agrarian society. Industrialisation and urbanisation brought about a gradual disruption. In the urban industrial setting the formal structure of the family has survived, but the role holders have lost the sense of togetherness and capacity for communication. Men have found themselves preoccupied with the role of breadwinner; women have experienced a sense of isolation and having to do nothing but carry out routine chores. Children have turned the home into a camp where they come to eat and sleep. They have their psychological and social location in other settings. This has led to the loss of identity of the family as a unit, and a sense of rootlessness and 'non-belongingness' among the individuals. Obviously this family structure is not congruent with the technological society. But, is there an alternative design?

The family performs many functions. It is the most important organisation for acculturation where beliefs, traditions, values and myths are internalised. It is a primary source for experiencing love, affection, receiving nurturance and acquiring the foundations of psychological security. It is a place where an individual can experiment with his impulses and learn to internalise their control and effective management. For the adult it provides psychological companionship, an opportunity to experience mutuality of concerns, and enables one to enjoy peace, privacy and togetherness. It is a space for repose. It is a system where interdependences of familial and social tasks are held in coherence. The family, in its dynamism embodies the fundamental meaning of social living.

The family still retains these functions, but it has ceased to be their sole repository. In a complex, technological society no single system can provide all significant meanings and sources of identities. For example, a woman today cannot derive her total meaning from her role as a housewife, wife or mother. In addition, most social institutions which traditionally permitted a woman (through rituals and other conventions) some outlets for her to experience sense of community belonging have disappeared. The only new activities are those of being a spectator, going window shopping or to clubs, which everybody cannot afford. Even such opportunities for being together like family meals, spending an evening in games, play or storytelling have largely vanished because of different time schedules of family members. The interdependencies of familial and social tasks have become fragmented in the family.

The family has to be redesigned for the technological era. For example, role divisions across age, sex, and hierarchy have to be done away with. Similarly overdefined sets of activities for each role have to be discarded. The family has to be structured around task, not role, differentiation. For example, the task of a breadwinner is a primary one. Originally it was exclusively the man's responsibility, but now women also perform the task. Most men, however, consider it a reflection of their inadequacy if their wives have to work. They see work for women only as an adjunct economic activity or an indulgence to keep them engaged. In our experience many women share this belief deep down. The fact that work is a source for developing potential and meaning is not taken into account. It is seen as merely an escape from the boredom and vanity of being only housewife, mother or wife. The technological society has created a change in the old pattern. And in the process of transition it has generated guilt, anxiety, resentment and conflicts. Even the kind of work to be done has been defined by sex. Men need challenging and hard physical or mental. Women need routine, mechanical work which needs patience but no challenge.

For example, cooking and kitchen work have been considered only fit for women. What stops men from joining in? Many men and women in our groups said that a working woman always had to come home and cook. The man who goes out to work comes home, and claims to be tired and exhausted. He refuses to help. Women seem to have more energy than men. Whatever few hours are available for being together are lost because the wife is occupied in the kitchen or is busy with the children. Men relax by themselves or socialise with friends.

A third area relates to nurturance and control. Women, especially, mothers, are supposed to provide nurturance and men control. Once again, the technological society has knocked out these role boundaries. Man is hardly available for control. This is more true in the wageearning families where the man spends long hours of the day at work. In effect, women have to provide nurturance as well as control. Man either reinforces social control and withholds nurturance as he has no time, or he pampers the children with excessive gifts but withholds love. The children experience a double bind. Their security is shaken and their sense of trust and wellbeing is lost. The participants from the second and third wave of the current generation gave a glimpse of the conflict in their interpersonal relations and intrapersonal relatedness. They communicated the view that their scepticism, self-centredness, disengagement from the system, and their feeling of disaffiliation originated in the experience of confusion and conflict in the family setting. These feelings were further reinforced through school and college experiences, but their genesis lay firmly in the family role structure.

The conflict and confusion about location of nurturance and control strengthened the root of male chauvinism and created a swing towards women's lib (female chauvinism) and thus has added to the battle of the sexes on the social scene. On other consequence at the intrapersonal level has been an increase in the confusion regarding differentiation between maleness and masculinity and femaleness and feminity. The confusion has created a transient culture of 'unisex' and/or a diluted cognitive culture of equality which only operates in public settings for display. One wonders why role holders in the family consider the activities of nurturance and control as being sex specific. These are processes for the child's wellbeing and acculturation and can be carried out jointly or by either parent.

Another function of the family is to help the child to learn to appraise reality and differentiate between feelings and impulses. In the role structured nuclear families a peculiar process has developed. The progenitors tended to protect the children from coming to terms with the economic, social and psychological reality of the family. Information was withheld in the name of protecting the children against the harshness of life. The progenitors considered the descendents either incapable of understanding, or too young to manage the realities of life. However, children invariably came to know of these realities from neighbours and friends. Most participants in our groups gave examples of how withholding discussion of the reality of the world destroyed their faith in the adults, induced a sense of worthlessness and shame, and blocked communication with the parents. It finally got them used to ignoring the appraisal of reality. The family discouraged the need to share problems together and encouraged the belief that 'everything is alright'. To impart training in reality appraisal, children need to be included. The membership of each individual in the system has to be affirmed by sharing. Trust, psychological security, and acceptance of reality can be fostered through this process which would also bring about more frankness among the members of the family. Their inclusion would foster familial interdependence and reestablish communication and togetherness.

The family process of covering up the reality, specially social and living reality, is also reflected at the national level. The government practice is to assure the public that 'everything is alright' and if there are unpleasant incidents, they are insignificant. This is often true about communal riots, strikes and other areas of civic conflict. Newspapers use bland statements such as 'violence broke out among two groups'. The groups are never named but localities are mentioned. Everybody knows who these two groups are. Figures supplied of the dead and injured are believed by what is cynically said is—'multiply the released figure by ten'. Participation in strikes, attendance at meetings and many such events are toned down or exaggerated. While the conflict is visible in many organisations, no official information is made available. Each individual is left to guess, scan, collect and establish his own truth. Rumours fly around and then people are cautioned against the evils of rumour mongering. One can understand the need for secrecy about defence and other sensitive information. But the need to gloss over those public events or events which are significant for social living and day-to-day transactions makes very little sense. It only undermines the sense of belonging to the system. It creates distrust of the authority in the system.

Finally, in the role-based structure of the family all critical decisions like the choice of subjects at school/college, career, and marriage are prerogatives of the progenitors. In a technological society this is not viable. It is not even possible because parents have little information or knowledge about courses, careers and potential professions. Participants in our study also gave examples of parents imposing on children choices they had been denied themselves. They often want to spare their children from suffering financial hardships and unpleasant experiences they themselves had suffered in their younger days. This tendency of overprotecting children from reality and pushing them into the role of receivers on the one hand and performers on the other hand denied them the discovery of their own personal strengths and potentials. Most participants had to stand up to the family after graduation because they did not want to act according to the choices imposed by the parents. Those who did not confront them directly tended to act on their own and present a fait accompli.

In an evolving society the individual cannot be a continuation of the culture, lifestyle and choices of the family. In these respects the family has to create a new beginning for each generation. It has to become a psychological convergence of individual identities but with a variety of life roles, styles and spaces. Replication of the life of the previous generation by each succeeding generation can only be crippling. Insistence on replication in the culture of transience is a hangover from the agrarian society.

The technological society and its ethos open up opportunities for each individual to begin afresh, to create a new heritage of his own through competence. The continuity of the family heritage needs to be re-examined. Does there have to be continuity in the contents of the heritage? Does it have to imply repetition of the same process? We believe that the continuity of the family heritage has to be anchored in the commitment to replenish the family ethos as a living force and move forward, rather than letting it become a compulsion to hug the past. The ethos and heritage of the family can still be carried forward and bettered if the institutionalisation of roles can be freed from age, sex and status differentiation.

The only and perhaps the most significant condition for designing the family as a relevant institution and an organisation for our era is to identify activity sets of the process of living together. These activity sets may have existed earlier but were grounded in the stratification of age, sex and status. In the technological society the activity sets have to be grounded in the time and location patterns of the living processes of the family members. They have to be carried out by the member available in the time slot regardless of sex and status. It may be a problem for the very young but by adolescence the membership needs to be initiated and individuals made responsible. Dissolution of differentiation by bio-social criteria is one critical step of the family in the technological ethos.

Thus, the undifferentiated hierarchical monolith of the family has to be dissolved. An organismic system, based on individual differentiation and potentials, has to be recreated through psychological integration and not through biological compulsions. The family has to move from a bio-social organisation to a socio-psychological institution.

Education as an Institution and Educational System as an Organisation

Let us take educational institutions as examples of another set of organisation. The family in the past was a monolith engaged in processes of primary socialisation. The education system on the other hand was an institution and an organisation for the processes of secondary socialisation. In the Indian tradition, the basic design of education was conceived as a means to dissolve the intensities of primary relatedness and introduce individuals to multiplicity and heterogeneity. This was meant to introduce the individual to the memberships of the community and society beyond the family. It encouraged investment and commitment in negotiated roles and tasks, helped to enhance individual choices and induct processes of selection and evaluation of living reality in order to make appropriate responses.

The education system today, being dependent on resources from the community, is open to forces from many directions. As such, its objectives become a matter of concern for the polity of the day and the consumers of its products, i.e., work organisations, parents, students and finally academicians. Each of these concerned parties looks at different aspects of education and builds pressures to influence the system which is therefore more vulnerable than the family.

The dependence of the education system for its resources increases its constituency to a membership which cannot be subsumed in a biosocial model. A socio-psychological model has to evolve. Ancient Indian society generally freed the education system from dependency for its material resources, by developing a unique land grant system and creating the ethos for a non-wage teacher identity. The teacher sought no wages but received gifts for his contribution. This provided the education system with its autonomy.

In the modern era the education system depends on resources from the state and individuals. The teachers have had to reduce their roles in order to avoid major confrontations with the polity and other resource controllers. In spite of having academic freedom and autonomy of governance, the dependence for resources has made it difficult for the education system to develop an operational integration of various roles within the system. This failure has recreated in the microcosm of the education system, the same dynamics that prevail in the macrocosm of society. Each role pulls the system in a different direction. The academicians insist on the purity of academic inputs and claim the sole right to design the contents of education. The academic convention of having only one person as 'professor–administrator'—who, like the administrative elite, acts as a know-all—frustrates the efforts of the committed teacher. The administrative and other staff pull the institution in yet another direction. Instead of accommodation and service, they believe in control. Their comparative marginality in terms of the objectives of education is caused by the teachers and the taught who see the existence of administrative staff only as a means for the satisfaction of their own needs. Delay, postponement, and living by the rules book are some of the methods used by the administrative staff to control and extend their power and counterbalance their marginality.

Students living through a period of psychological moratorium see the education system as a context in which to be and to become. Confronted with rigid syllabi and limited availability of infrastructural facilities for their psychological moratorium, they create a culture of indifference, escapade and protest. They learn to make minimum efforts for maximal returns and reduce the task of learning to meaningless and compulsive boredom. Their demand for making education relevant has become a slogan. Their demand for building infrastructures for help in the psychological moratorium is perceived as a threat by the teachers.

Consequently, that part of the education system designed for mutual learning has been reduced to two mutually exclusive roles, the teacher and the taught. The teacher is the giver of knowledge, skills and attitudes. His job is to transform the novice from the raw material stage to a finished product stage for use by organisations. The roles of the taught is to receive unquestioned what is given. Growth is measured in terms of the ability to reproduce what is taught. The interaction between the students and teachers is largely restricted to classrooms. There is little space for intellectual dialogue between the two. The purpose of education has been reduced to information dissemination and generation of knowledge largely theoretical in nature. Very little thought is given to its application.

The participants did not experience the education system as one providing enthusiasm, inspiration and freedom. They did not feel impelled to discover new horizons of the world and the human existence, with its accompanying intellectual ferment. They experienced it as dull, indifferent, routine and an enforced postponement of their role taking in society. To quote one of our participants, 'Today's education, restricted as it is to information and acquisition of techniques, is no more a mind-blowing thing.' It is merely a task 'which has to be done'. A large number of our participants agreed that the education system had not challenged their potential. They could cope with its demands by giving it only two months of their time in the entire academic year. This was sufficient to achieve the level of academic performance which was socially acceptable and which ensured openings either in higher education or in the job market. They also claimed that enlightenment came to them from their peers or guest speakers. The contribution of their own teachers was rather low.

As we sat listening to the accounts given by a whole set of young students and managers, of their experiences with the education system, we were struck by the following points:

1. All institutional and infrastructural processes have been eroded. Education has become another task of production. In terms of organisation processes, it has regressed into a rigid hierarchical organisation with all authority centralised in one person. These processes are very similar to the processes in an agrarian society. All subsystems of education are managed by one person. Each department can have only one professor. He is the administrative head and a technical expert. He controls all the resources. He has to be addressed as 'Sir' by the lecturers and readers in the department. He does not have peers in his own domain. Each subsystem is encapsulated and isolated. There is hardly any transaction across departments in the common endeavour of education. This is not only true of the teachers, but also of the taught.
2. The educational institutions are preoccupied with maintaining their status quo. They struggle to maintain the stability of their structures. They have also overcrystallised the role hierarchy and specialisation. Responsiveness to the changing reality of the environment, a necessary quality of any education system, has been lost. Maintaining role differentiations and academic freedom has acquired centrality. The current structure and ethos have marginalised education tasks. Just as the administrative elite has learnt to live in administrative reality delinked from the living reality, so has the educational elite learnt to live in an academic reality—of cause–effect analysis, post-event explanations, and theoretical construction through abstract knowledge.
3. In the minds of academicians, administrators and polity, the taught are held in the role of 'kids' who know 'no better' and who definitely 'do not know what is good for them'. They are perceived as highly susceptible and lacking discrimination. They are held in suspicion and perceived as wasting their time, indulging in non-essentials, and avoiding hard work. They have become

objects of sermonising. However, the academia itself is above criticism and self-righteous about its academic decisions.

These three aspects of education today have turned the system into a factory. It has lost its status as a significant institution of secondary socialisation. The objectives of dissolving the intensities of primary cathexes, of introducing multiplicity and heterogeneity in object relatedness and other aspects have been lost sight of. The participants attributed to education a propensity to create indifferences, scepticism and isolationism. In the midst of multiple and heterogeneous population entrenchments, they emerged as against negotiated commitment. Individuals learnt to hide the living reality rather than explore it.

The objectives of secondary socialisation as stated earlier were postulated for the era when society was predominantly agrarian. They are still viable in terms of the age group experiencing an enforced moratorium leading to maturity. Earlier they could be managed by the ethos of obedience, respect, conformity and advice. They could also be managed by a set of do's and don'ts. Today these same processes have reproduced, like everywhere else in society, the ethos of control and coordination. The participants often stated how the personalised do's and don'ts of the parents and significant surrogates from the primary system were extended into the impersonal rules and regulations of the secondary system, with the common element of domination through coordination and control. While the family provides space for expression of pathos, the education system denies such space.

The education systems have to discover processes related to the emergent learning ethos and its relationships. The ethos based on building equations rather than demanding obedience, on defining the assumptions of rules and regulations rather than issuing them as commands, and on joint reviews of purposes of togetherness in education rather than stating them unilaterally and demanding conformity. To build a new India, educational institutions have to be conceived primarily as settings of growth for human beings and not primarily as producing technical experts and employees. Education in India and in similar societies must include a large segment of experience-based learning, as was the case in the *Upanishadic* tradition. Exposure to the technology of social living and social action, and clarification of the underlying assumptions, must become a significant component of the education system, along with training in technology of production, theory and scientific enquiry. Education in India and other similar societies has to add relevant value dimensions to its earlier objectives. Some of the salient objectives we gathered from the participants were fostering of voluntary relatedness, and creation and replenishment of community life spaces and attitudes of social and psychological competence, together with occupational competence. The introduction and promotion of psycho-cultural and psycho-philosophical analyses, in order to build perspectives of knowledge and its applications, became another objective.

Consequently, we suggest that the construct of academic autonomy and the value of free education have to be re-examined. A way has to be found to raise resources for education so that it is free from the excessive control of administrative and political processes. A new structure of managing the education systems, where resource contributors do not have the final management responsibility, has to be devised.

All the above suggestions may be felt to be impractical. However, we are stating the inferences reached by the participants from their own experiences. According to them the education system cannot abrogate its wider responsibility for individual growth in the name of efficient imparting of knowledge. As with the institution of the family, the stability of educational institutions in the new era must lie in the commitment to the growth of all participants. Learning would have to become a process of cognitive, emotive and action integration.

Simultaneity of Socialisation and Individuation

Socialisation is an accepted task of society. A child's instinctual energies have to be channelised into socially acceptable forms of behaviour. Socialisation, then, is a mutual task of transformation, inclusion, and emergence of role definitions for individuals. According to our data from the participants of the second and third waves of the current generation, an exclusive emphasis on achievement during socialisation made them feel guilty about their own being. It created a sense of deprivation and denial. Inconsistency in the context of socialisation resulted in a sense of confusion and discrimination. Their experience of socialisation was coloured with a sense of rejection. It created a compulsion to conform in action, while withholding their feelings. The seeds of alienation from the system and estrangement from the self were sown during this process.

Socialisation lays the primary foundations of security, trust and belonging. This function has to remain. The child's needs of dependency on and control, guidance and modelling by the adult world cannot be ignored and must be attended to. Without these basic foundations, a growing child will become an anxious, tense and restless individual, with no source of stability, direction or integrity. Socialisation also provides prototype models for adulthood roles. However, the primary system tends to leave no space for the child to be an individual. In laying down the basic foundations of identity, the primary system treats the child as an extension, an echo, and a shadow of the parents. Many parents in the primary system forget that every child is 'twice-born'. A child is born into the family as a son or a daughter, and is reborn into the community as a member in his or her own right. Parents concentrate on the task of shaping their child into a proper son or a daughter and totally neglect that of grooming the child as a member of the community. Such an attitude tends to stamp out the individuality, initiative and autonomy of the child. Even as the child grows up, the parents continue to treat him as dependent and immature. They block the development of sentient and task interdependence.

We suggest that the parents' attitude results in harrowing experiences during the process of growing up. The participants in the group experienced the dilemma of retaining their son's or daughter's role even on becoming members of society. In most instances, they failed to be reborn as members of the community and ended up as its consumers. They tended to use the community for their own enhancement and at best for their family's. They failed to make a healthy commitment to society at large. Is it not obvious, then, that the very nature of the socialisation process in the Indian family is a major source of failure as an instrument of social change in India? Social change also requires commitment to society at large.

The parent's interpretation of the task of socialisation tends to create an overdefined concept of individuals. They develop a strong need to remain confined within prescribed boundaries by denying most of the needs, feelings and wishes that they experience as persons. They tend to be afraid to question the system. The fear is backed by looking at their status in the system only in terms of doing what is required it. The fear makes them deny their role as representatives who can act with the system. Eventually, this entrenchment pushes the individuals into prizing the secondary gains achieved by the denial of the self. They start by not only resisting change, but also by viewing all change as chaos, which will deprive them of their gains. They become withholders rather than participants in the system. Finally, it undermines their ability to counteract the dehumanising processes of the system. The alternative is to disaffiliate oneself in feelings, alienate oneself in thought and rebel in action. The process encourages indifference, scepticism and finally disengagement. Replenishment of the system and simultaneity of action for the well-being of the self and system are lost.

The task of the modern society goes far beyond that of traditional socialisation. The modern society needs to institutionalise the process of individuation. Individuation means an individual acquiring the ability and the attitudes for evaluating the system openly and stating his experiences and inferences. This includes being able to stand up to such processes of the system that tend to destroy representative membership. He should be able to aspire for dignity in a role he performs without restricting himself to a prescriptive model. Individuation implies the ability to validate and affirm oneself in realistic terms and not disowning oneself in the light of invalidation from outside. A new Indian society will not emerge unless its members are willing to accept validation from within as an important anchor of the self and the system.

Unfortunately, the characteristics of individuation in society, which are important for creating transactional situations, have come to be regarded as rebellions. We have already stated that, structurally, the new society has to have separate but simultaneous anchors of stability and change. We are now suggesting that the individual of the society-to-be should have the strengths of primary socialisation from the family and the strengths of individuation through secondary socialisation from educational institutions. Without these dual strengths, the individual cannot turn his potential into a constructive force for society's growth. An individual today needs these strengths in order to transcend the role of a son or a daughter and to act as an active member of the community.

No society of the future, with its vastly complex and differentiated subsystems impinging upon the individual from all directions, can afford to provide only traditional socialisation. This produces, in agrarian society in general and particularly in India, the unidimensional role of a son or a daughter. It is backed by a social organisation where all secondary systems are mere replicative elaborations of the primary system. They jointly foster conformity, compliance and system dependence as primary values for the individual. This set of primary values only creates mechanical human beings and dictatorial empires in the name of managing a social system. The need of the society-to-be is to have individuals who can relate to multiple systems in multiple roles with differentiated attitudes without losing their sense of identity or personhood. This is only possible through individuation, which has to be fostered by the education system. Parents have to recognise that with each spurt of growth of their child, they need to restructure their role transactions. They have to keep in focus their own and the child's dual roles, of being parents/children on the one hand, and members of the same community as equals on the other. Unless they recognise the duality of the roles, they will continue to create in the child a sense of guilt towards growth.

In agrarian societies secondary socialisation reinforced the attitudes, values and life orientations of primary socialisation. In traditional societies there was no need for integration as socialisation was a continuum of values and processes from the microcosm of the family to the macrocosm of society. Each subsystem was contained within a larger system. Attitudes, behaviour and orientations of the primary system forged the essential framework for roles in all the systems of society. The technological society intervenes in a big way to create discontinuity between the primary system (and its socialisation) and secondary systems. Family, education and work, the three major institutions of society, demand a discontinuity in attitudes and processes of roles. The recognition of the fact that the larger society of today is not a continuous and homogeneous unfolding of the family, but a complex structure of discrete and differential systems each demanding different roles from the individual, is essential. The individual needs to shift the locus of his identity and acquire attitudes and orientations to be able to hold multiple roles simultaneously. To achieve this, both primary and secondary socialisation in the family and secondary socialisation in the education system have to provide inputs for individuation. The processes of socialisation in the two cannot be identical; they have to be distinct but integrative in the individual. Integration could be achieved by designing an institutional infrastructure within and between the family and education systems.

The tradition of agrarian society has only one model of socialisation, the bio-social model. This model does not envisage drastic discontinuities between being a member of the family and a member of the community, as the technological society model tends to demand. The goal of socialisation in the agrarian society was to initiate the individual in a role identity. It did not take into account the shift from dependence on others and interdependence with other roles, to dependence on self and interdependence in tasks with others. The technological society expects the individuals to make this shift and overhaul his identity as he moves from the family to the community. It demands personhood, not role identity. The role integration that the individual achieves as a child cannot serve him for the rest of his life. He has to achieve role integration anew, at least once during his young adulthood, if not repeatedly throughout his life.

The discontinuity between familial socialisation and community socialisation is a fact of life. The period of maturity is extended beyond the family to the work community today. The failure to recognise the discontinuity and consequent postponement of the development of the person has led the individual into a wasteland. The individual feels rootless, anxious and experiences a sense of non-belonging. He feels alienated from the family and the education system. He resorts to a self-centred ethos. To counteract alienation, the individual needs to internalise the process of individuation. Alienation, basically reactive in its origin, creates non-committed individuals. The individual's evaluation of the system is over-focussed on its weaknesses and hypocrisies. He ignores the strengths of society. It makes the individual brittle and fragile. His criticism is tinged with cynicism. The social system therefore learns to ignore, suppress or, at best, preach to him. Sometimes he develops the 'outsider' syndrome. He may free himself from the social system, but he does not necessarily find the strength to contribute to the reconstruction of society.

The current society's mode of dealing with alienation is also reactive. It harangues the alienated and refuses to have a critical look at itself. We suggest that society should shift the focus of secondary socialisation from reinforcement of the role definition of the primary system to the institutionalisation of a serious, constructive appraisal of values, attitudes and the current realities of the life space. This would begin the processes of individuation without the loss of identity.

Education for Continuing Self-renewal and Role Change

The assumption of the agrarian society that the individual is all set for his life after initial socialisation has to be discarded by the technological society. Any role or self-integration, i.e., the identity, that an individual achieves at any given point during his life will at some point become obsolete. Self-renewal and regeneration, redefinition and restructuring of role or, in other words, achieving a new integration of self and role are essential processes of human existence in a technological society. Self-renewal and regeneration involve discovering new goals, new directions and new meanings congruent with the changing interface between the environment and the life space. Self-renewal and regeneration can help the individual forge a new relatedness of the self with the role and system. The ability to modify the roles, and attitudes underlying the roles, can lead the individual to counteract the absolutism of the system and the role to which the system tends to bind him.

There is an urgent need for redesigning the network of attitudes in forming social roles in order to create a technological India. Our data has very pointedly brought out the failure of parents, teachers and other significant actors in redefining and restructuring their own roles. It has been the most annihilating aspect of the youths' lives. Parents and educators are unable or unwilling to recognise that, as controllers, they are responsible for the scepticism of the young about the self and the system.

In a large number of marriage counselling sessions we have encountered the same syndrome. Husbands and wives have been blind towards the need for change in the quality of relationships and activity sets at various stages of their married lives. They have held on to the same role definition and the accompanying attitudes they had when they entered marriage. Utter insensitivity to growth—both psychological and psycho-biological—has become a well-entrenched feature. We see the same problem in society at large. The failure of developmental tasks in India is another analogue of primary relationships. The administrative cadre, it seems, has failed to redefine and restructure the traditional roles of controller of resources, relief giver to crisis victims, and upholder of law, with the result that developmental processes have not taken root among the masses. The task is still performed with the same approach as the other activities of governance.

It is clear that the skills to review, renew and regenerate roles for role–self integrations cannot be completely acquired by the child during his primary socialisation. Home is too small a system for an individual to discover the multiplicity of self and roles. The individuals who can contribute to the growth of a new culture in India must acquire the strength to confront the invalidation of the role and the system without fear of losing the meaning of life. The education system has to accept greater responsibility for inculcating the strengths and attitudes needed for self-renewal. This can only be done by institutionalising the process of individuation and reinforcing it through creating opportunities for self-renewal.

The ever changing technology of production, transportation and communication has acquired a momentum which continues to transform the quality of transactions of the individual regardless of the given social structure. It is necessary, therefore, that the technology of social life keep pace with these changes. In order to synchronise the movement of these two types of technologies, both the individual as well as the system have to respond quickly and effectively to the widening gap between the emergent transactions and the role expectations of an individual in a system. Unless this is done, the individual will develop feelings of guilt. Technology has opened new opportunities for young men and women to interact without an escort. It has led to a new overt pattern of behaviour between them. It has created opportunities for voluntary relationships. The basic role conceptions and attitudes learnt from primary socialisation tend to clash with these opportunities. This has created a sense of guilt, anxiety and other self-condemning patterns. Development of the technology of social living is essential to keep pace with other technological changes. To us, the key to such development lies in the education system.

We have identified what we consider to be the three basic anchor processes which can bring about a new society in India. If these processes can be institutionalised, individuals will find the conviction and courage to participate in the growth of society. The participants in groups, while agreeing with the identification of the three processes as potentially capable of beginning a new society, were not in complete agreement with us.

Perhaps the parents and educators in the current system would consider these processes as leading to chaos. The reaction to these processes has been like the Queen of Hearts who cries, 'Verdict first, evidence after.' Cognitively, most individuals want to bring about change. However, their need for control and fear of their own invalidation continues to make such individuals act from the stance of patronage. They assume that they alone know what is good for the individual and society.

When we presented the data from the individuals and groups to the significant roles (parents and educators), they responded by saying that this was not real data. When we persisted, they became defensive. They made statements like, 'How can these kids think like this?', 'We have done so much', 'These people do not understand the reality of the society', and finally 'What you propose will create chaos.' Examples of chaos in society are Naxalism, the Nav Nirman movement of Gujarat, and scores of other such disturbances we are witnessing all around. These are also a syndrome of moral decay. These role holders have developed two other stereotyped explanations for these disturbances. According to some these are expressions of uneven distribution of economic resources, unemployed youth and the rural–urban drift. All these explanations have been flogged to death. Even if they were true, the fact that there is chaos cannot be denied. The current chaos has also been created by the rigid stance of policy-makers in upholding yesterday's role conceptions of a traditional society. They believe in hierarchy and a one-person system. The policy-makers believe that the current stress is only a product of circumstances beyond their control. However, the current chaos is an inevitable result of the policy-makers having turned a blind eye to the living reality of today's society.

We could ask yet another question: What about the psychic chaos the concept of traditional roles has created in the lives of young men and women? Traditional attitudes and orientations to society as a system have reduced young men and women to non-beings who live by mere reactivity. The denial of membership to the current generation has destroyed their ability to replenish the system. It is so upsetting to see how these come forward for social reconstruction being given without a chance to state their own point of view. Holding on to power processes and role modalities of the past contribute to the morbidity of human existence. How long will the planners continue to wreak destruction and humiliate human beings? The concern of the present, in the here and now, is a search for an alternative to the reactive mass violence and disruptive forces that are appearing on the Indian scene. Is it possible that the country's precious heritages can be saved and a new society reformulated? We record our explorations of this question in the next section.

The Alternatives

Once upon a time the complexity of multilevel, multifaced and sentient interdependent relationships characterised Indian society. Its technology of production was simple. It was an integrated and cohesive society and a culture, though not a nation. The culture of transience has undermined the old complexity and created a different kind of complexity of socio-economic and socio-cultural transactions. Exposed to innumerable stimuli from a number of sources simultaneously, society has continued to borrow divergent forms and sometimes processes. Today it is caught up in having to choose from among a variety of forms and processes in order to bring about cohesion and integration. Massive efforts are needed for its reconstruction, requiring the institutionalisation of a host of new processes. Reinterpretation of values, realignment of goals, and development of appropriate attitudes are required to make Indian society dynamic. Changes in the basic processes of relatedness between man and the system are needed urgently.

In order to identify these processes, a relevant concept of man and his nature, a concept of collectivity and the nature of relatedness between them have to be identified and chosen. On the Indian scene today multiple concepts of man, collectivity and their relatedness, contradictory to and contrasting with each other, operate. This creates confusion in choosing relevant processes for convergence and coherence.

In our discussions we discovered that the size of the task as well as the levels of choices involved are experienced by people only cognitively. For action they see this task fit only for a messiah or a new avtar.

The messiahs known in human history have operated in simple societies characterised by overstructured and overpolarised roles, values and rituals. People were close to their experience. It was only necessary for the messiah to invoke faith. His task was to recreate a dignified niche for man in the system by confronting authority persistently, sometimes at the cost of his own life. The messiah could set an example for realigning people and the system. The question is whether the messiah would succeed in his task today? Let us take the example of Gandhi—a truly charismatic individual endowed with the qualities and the vision of a messiah. He could mobilise the masses for a coordinated and concerted action to achieve the well-defined goal of political independence. However, he could not mobilise many to participate in his social reconstruction programmes. He already had a blueprint for a new Indian society. But the majority of the Western educated Indian elite could not respond to him and his ideas in this sphere. Those who did, such as the Sarvodaya workers, could not free themselves from the basic agrarian ethos of India—an ethos which had already become stagnant.

Even if we accept the possibility of a messiah coming to wrestle successfully with the Indian ethos, the basic problem is whether he can mobilise people into dynamic and innovative action, so as to enable the continuous unfolding of society, system and individuals. In our review we found messiahs who injected dynamism but failed to institutionalise the process of self-sustaining change. As such, most messiahs generated newer models of meeting life situations and created newer definitions of man and his collectivity. They also created new choices. However, they failed to create processes for internalising the principle of change in the structure of the society. Their modes of meeting life situations, their definitions and new choices became limited and frozen. Their disciples followed in their footsteps and ended up being mere shadows. In the midst of existential struggle they surrendered at the feet of the master, conforming, obeying and repeating. A messiah by his very nature creates a gulf between himself and his followers which becomes difficult to bridge. They, thus, acquire techniques of becoming, but fail to remain a Being within.

A messiah may reconstruct Indian society, but there is always a possibility that society may not become dynamic. It is doubtful if a messiah can generate enough dynamism to shake off the inertia born out of the containment and detachment which has characterised society for over a thousand years. The last four decades have shown that a charismatic leader leaves an organisation mortgaged to himself. It starts to disintegrate upon his death. We know of no example where the charismatic leader has been able to institutionalise his own processes into the dynamic ethos of an organisation. At best, he has produced a model to be followed. We are reminded of an Indian saying, *'Haveli ki umar sath saal'* (The life of a palace is sixty years). A 'house' created by a dynamic person lasts only three generations. By the middle of the third generation it begins to decay, losing its dynamic and integrative forces. In fact this epigram appears applicable to all agrarian societies in Asia.[1] The synergy generated by the charismatic leader is dissipated by the third generation. After the death of the charismatic leader the organisation—be it family or business—becomes involved in property disputes at the worst, or with maintenance without deployment or generation of further resources at best.

History has shown how the new synergy released by the messiah creates a new ethos, which by the third generation becomes binding on people. Breakaway groups adopt different aspects of the ethos as dogmas conflict with each other. In the ensuing struggle an individual looses his locus. The system becomes unmanageable.

As such, Indian culture and history suggest that a secular charismatic leader might actually prove dangerous for society. As against a messiah using personal trust to win people, a charismatic leader uses reason and logic. The followers of a charismatic leader take this logic as sacrosanct in the same way as the followers of a messiah take his utterances. Like the messiahs, the charismatic leader also leaves a vacuum after his death. His place is always filled by conformists who are committed to creating a rigid and stable universe on the lines laid down by their master. They are set on completing this task and at best modifying it, but never innovate. The rifts between the followers on the issue of who will inherit the mantle have torn many a movements asunder. The history of communism validates our statement about the secular charismatic leader. When the followers of a charismatic leader have failed to respond to changing conditions, the society has been taken over by another strong person and the work of the earlier charismatic leader is wiped out. Military dictatorships in the Third World are arch examples of this.

A charismatic leader or a messiah can only succeed by triggering off the dynamics of reason or faith. If he uses the dynamics of reason, he blocks the eventual unfolding of the individual's own self. The followers learn to live for him and his system, perpetuating the traditional processes. Can Indian society afford repeated changes in form without changes in the basic processes? Our answer is an emphatic 'No'. In the absence of a messiah or a charismatic leader, can the dominant role holders of the current social system collectively do

Pearl S. Buck, in her novel *Good Earth*, exemplifies it in the Chinese set-up.

the required tasks? The probability of this happening seems to be low.

Scepticism, the basic nihilism of the Indian youth, prevents the ushering in of a new society. It manifests itself in a range of slogans or aphorisms, varying from mild criticism and passive attitudes, through a strong negative orientation and spectator role in action, to utter cynicism displayed in thought, feeling and action. Even as an individual prepares to give up his negative stance, the phantoms of rejection, dependence, control, deprivation, loneliness, isolation and guilt rise up to chase the individual back into his safe sceptical role.

Given the this state, neither the faith of a messiah nor the dynamics of reason of a charismatic leader can change people for enduring social action. The dynamics of faith alone can only create disciples of maharishis, *gurudevas* (reverand teachers) acharyas and *babas* (ascetics). It can perhaps be utilised for personal solace by an individual when the stress of existential reality reaches the nadir of ennui. Faith, to rephrase Marx, is only an opium for non-action. The dynamics of reason alone can only make polemic writers, public speakers and political pundits who, when off-stage, sink back into the swamp of exploitative existence.

Attitudinal ambivalence—'I am, I am not', 'I can, I cannot', and 'I do, I do not'—is deeply ingrained in the individuals of today and imprisons them in non-action. They are dissatisfied with the processes of the society but when the time for action comes, they end up fighting for the forms. Many movements in India are classic examples of this tendency. Individuals organised themselves to bring about changes in the processes, but ended up fighting the structures. Scepticism and ambivalence in our opinion leave little scope for any effective action. These feelings must first be dissolved and this can happen only from within.

The Ethos of Criticism in Action

It seems, therefore, that the choice before the individual is either to remain as he is or to become his own messiah. We believe that it is for each of us to commit ourselves to a responsive style of life in the microcosm of our immediate psycho-social existence—the family, place of work, immediate neighbourhood and community. When we posed this alternative to the participants, they were disappointed and disturbed. They believed that their initial task was to generate social criticism and spread awareness and concern among a wider set; the action would automatically follow. They considered responsiveness to their own life space a secondary step.

Modern society is like a sponge soaking up verbal social criticism. Its power structure can withstand the pressure of various groups, by labelling them fascists, militants, terrorists, reactionaries and communists. It can manage and manipulate opinions effectively. People see no option but to form groups to take direct action. The power structure suppresses such groups ruthlessly. What appears as a surrender by the power structure is mere eye-wash. From every encounter with such a group it derives, like Bali of the *Ramayana*, redoubled strength to continue its own processes. We, therefore, believe that to build healthy social criticism and to generate action via criticism is a retrograde step. Individuals who take this line would either end up as dilettantes or become absorbed into the political belief system. Modern society can only be tackled by individual criticism in action. Unless individuals recognise this, they will continue to be party to the social and moral crime of giving up on themselves and the system. Then the only option left is violence which is destructive to the individual, the system and the nation.

We suggest to individuals the path of personal criticism in action. But what does it mean? It means that each has to act for himself in his own microcosm. If he forgets this, he is likely to become a part of a movement, a member of a group or a cult. A movement, a group or a cult in a modern society demands more conformity than they did in a traditional society. Without such conformity the movement becomes a victim of fissiparous tendencies. The group can tolerate no divergence. The realities of society becomes confined in an overstructured framework. Hence the primary need for today's generation is to end this dependence on a group and its validation by a majority. To stand alone and to act for oneself in one's socio-psychological space is the call of the day. It is the only alternative.

So we are back to where we began. The individual himself is the crux of his problem. To quote Aldrich Cleaver, 'If you are not a part of the solution, you are a part of the problem.' Individuals in society will have to stop being a part of the problem and begin to be part of the solution. It is not a solution for society but a solution for his own existential and socio-psychological space. The suggestion to offer criticism in action to society is not new. Individuals throughout human history have acted in this mode. Some of them like Christ and Buddha, Joan of Arc and Gandhi created movements which in themselves offered criticism in action. However, the ethos of offering criticism in action is so threatening to society and those in positions of power that such individuals sooner or later become objects of reactive feelings. They are then treated as outcasts, criminals and martyrs; and in their persistence are finally deified. In India most of them were confined to pockets of society as initiators of cults. The ethos of offering criticism in action is never allowed to be institutionalised in society. It is turned into a tight structure. The failure of such people has been due to the fact that they chose macro issues to offer criticism in action.

Consequently, offering criticism in action has become a rather lonely struggle, requiring fortitude and commitment. The new generation has to compare the loneliness of offering criticism in action to the loneliness and estrangement they experience within. Acting out of the loneliness and estrangement, with its ambiguities, may prove to be healthier than the morbid anxieties and insecurities that keep the self in its present mode of social existence.

This path would free the individual from ritualistic conformity in his interpersonal behaviour. For example, he may recognise that respect for elders does not demand silent, grumbling compliance. It could mean a discussion of issues where the elder's dignity is included in the youth's honest concern for reaching an understanding. A son's rejection of the bride chosen by his parents may not mean disrespect or disobedience. It can be proof of the discovery that besides being a son, he is also an individual. It is an act denying the following assumptions of agrarian society: that a son is an extension of the parents and the family; that he has no separate, distinct identity; and that he should never act for himself but only for the family. Accordingly, the realisation that growth is a process of continual differentiation from and reintegration with the family and society has to be faced. From the data it is obvious that parents throttle the basic process of differentiation of an individual in childhood. The child's attempts be different and individuate are seen as disobedience and disrespect; or such attempts are considered impetuous outbreaks which need to be controlled and tamed. We understand parents themselves experience difficulty in growing with the child and creating reintegration of relatedness. The individual today has to recognise that the cries of disrespect, disobedience, dissent and deviance and remnants of the parents' entrenchment in the agrarian ethos and their own inability to redefine their own roles.

Creating an ethos of criticism in action is a significant alternative. What does this alternative involve? It involves discovering a new concept of a social man which transcends the assumptions of an agrarian ethos.

Concept of Man Beyond the Role

As we have seen, the agrarian concept of man has been management of the self in role or roles. In this assumption role or roles are the only legitimate foci for evaluation, assessment and transaction. Any impingement of the self on the role which generates role inappropriate behaviour, feeling and thought is suspect and censored. The new concept would have to accept the self as the coordinating agent of role or roles. The concept demands socialisation in managing the roles of the self. The shift also implies dissolution of normative absolutism of behaviour and recognition of the changing contexts of transactions. Participants in transactions would have to negotiate the context of their transactional situation and its realities rather than wrestle with expectations in the role.

The process implies that the individual has to dissolve his 'me-self', which is a product of oversocialisation. The 'me-self' is a product of cumulative inferences about the self and its nature received from the system and its significant roles. It is the crystallised and validated self. It is the basis for becoming an echo, a shadow. One of the primary assumptions of the me-self is the acted-upon status of an individual in the system. His role is to live by appropriate role feelings. The me-self, then, is a major foundation of the role society and the agrarian ethos.

The new concept of man, necessary for the growth of a technological ethos and the society beyond it, has to be anchored in the 'I-self'. It begins with the assumption that the individual has also the status of acting-upon in the system. It is his primary status and needs to be fostered. His status of being the acted-upon during childhood has to be counterbalanced with the induction of the processes of individuation. The process would involve simultaneity of the self and the system and not the system above the individual.

Role, in this model, is only an interface between individuals who constitute the system. The hypothesis of self as the coordinating agent demands that roles and the interface between individuals can be defined and redefined to foster the wellbeing of individuals and as such the system. The individual in his I-self accepts the fact that the situation and the task define the boundaries of the specific roles for the time and for the place the individual is in. But this role is not the totality of his self. It is also not defined by present role acts. He has the choice to determine his role acts.

Let us examine the implications of me-self as a concept of self promoted by the agrarian ethos. The me-self makes consistency in beliefs, attitudes and behaviour choices across situations, people and time almost sacrosanct. This imprisons the individual in a monorole. His role universe revolves exclusively around a single primary relationship. All other relationships are either secondary or analogous extensions of the main relationship. He lives in a unidimensional and unidirectional life space. The me-self, then, surrenders itself to prescriptive role acts and becomes a victim of its absolutism. Oversocialisation into a role and surrender of personal responsiveness to the situation follow inevitably. In our view, socialisation of modern man has to be matched with training in individuation. Individuation alone can counteract the forces of the system dragging the individual towards me-self. It is a relevant alternative to the current malaise of alienation. Individuation as a processes implies fostering a commitment to simultaneity of the self and the system. It gives man the courage to question and transcend the illusion of alternatives created by the role society.

Modern man needs to recognise the I-am-that-and-more assumption as an integral part of the concept of man. The dynamism of I-am-that-and-more opens up the individual for a role change, for self-renewal and for self-regeneration. It is the source of continuous social creativity. It is also a challenge for growth. I-am-that-and-more definition is the central theme of individuation. In it lies the source of a multiplicity of roles whereby the self derives meaning from itself rather than from prescriptive roles. It provides a counterbalance to validation from outside. It becomes the source of validation from within.

Accepting such a definition of self, the individual will find it easy to accept simultaneous membership of multiple groups and systems. It will be easy to say I am the son and more, and it is my 'moreness' that the family has to experience and accept. It will be easy then for the individual to be in the role of a son and also be a responsive participant in the family system. He will find a way to define the system. He will find the courage to persist, the courage to differ and yet be respected. In terms of growth it will imply a primary differentiation from all systems of belonging, and then through active choice, a reintegration with them.

The agrarian society forces the individual into relationship networks. All relations are defined in an apex hierarchy of superior and subordinates. The agrarian society systematises and crystallises the role acts and expectations into permanently fixed patterns. The I-am-that-and-more assumption of the nature of self recognises stagnation in the patterns of relationships and the need for their continual renewal. Bringing the 'moreness' of the self into a role builds a counterpoise to the fossilisation of role-boundedness. This assumption should be the basis for recognition by parents and individuals of the twice-born status of the child. In essence, this assumption brings into play the natural spontaneity, zest and dynamism of the self for the benefit of society. It can become a source of an evolving process of the self and the system simultaneously.

The Assumption of Criticism in Action

This assumption can be summarised in the statement: I have no certain knowledge but I have no doubts. In order to sustain the acting-upon status of the self, the individual has to learn to act with a certain degree of risk. He cannot wait to establish absolute certainty or the truth, the whole truth and nothing but the truth for action. He has to learn to work with approximations. In the moment of action the individual does not have doubts about what he is doing, though he may be acutely aware of the limitations under which he is operating. He works continuously and overcomes limitations in action. Each action is then an experience to broaden his horizons. This is the pragmatic principle of experience-based learning.

The agrarian ethos prescribes role acts, role appropriate feelings and thoughts. An individual's feelings, thoughts and fantasies at the level of the self were considered irrelevant. The agrarian ethos thus overloads the individual with residual feelings. These are held in abeyance according to the philosophy of detachment, sacrifice and self-negation. When such a society reaches the point where the individual loses the primary gains of the surrender of the self to the system, as has happened in Indian society, everything either seems futile or becomes an object of acquisition and need fulfilment.

The action definition of the self—I have no certain knowledge but I have no doubts—opens new horizons as a significant basis of the concept of self. It provides strength for experimenting with actions beyond the established modes of meeting life situations. We see in this definition the possibility of breaking through the cultural inhibitions against innovative action. It saves an individual from being smothered by bits of advice such as, 'Know before you act, 'Ensure before you act', 'Look before you leap', and 'Let us examine the pros and cons.' In themselves these adages are not bad. They caution the individual. But they are used by parents and others in the system to postpone action on central issues. They are used by them to create doubts and by implication reflect adversely on a person's maturity. Consequently the individual is excluded from effective participation in decision-making.

Commitment to the self and the system cannot develop unless the individual is willing to accept the responsibility for actions, despite doubts. Most individuals shirk this commitment because they are afraid of failure, and also because it means accepting full responsibility for their actions; the individual has to be his own agent and not a medium through which actions occur. The deepest form of internalisation of the agrarian ethos is reflected in Indian languages and their syntax which is primarily passive. The self is postulated as a medium of action and not as an agent of action. This suggests that the new commitment will require a modified syntax.

The assumption—I have no certain knowledge but I have no doubts—can free the individual to process information in terms of the present and contextual commitments rather than remain mortgaged to past commitments. It can generate the courage to admit failure without fear of invalidation. Learning becomes a process of generating wisdom from experience rather than knowledge from the given abstractions of the system. Experiential review of the self and role becomes a significant part of the repertoire of the individual. Tradition becomes a point to begin from and not a brake on action.

In our view these two concepts of man and action can generate the processes of being and becoming simultaneously. In the current Indian society the process of being and becoming are seen as mutually exclusive. There was simultaneity of being and becoming in the classical ethos of India unlike in agrarian society. Processes of becoming lead the individual into chasing social rewards. They generate the usual consumption and self-oriented patterns of unfolding of the me-self. The processes of being in the current Indian society have become synonymous with the idealised norm of being detached from life and making sacrifices. The first leads to glorification and deification, the second to fatalism and martyrdom. They do not synchronise to make a person human. Our need today is not for deities. We need human beings.

When these two assumptions form the concept of self, the individual will learn to define his relationships with the collectivity in terms of mutuality and replenishment. He will lose his fear of the collectivity, overcome his helplessness, and learn the skill of defining. He will lose his fear of loneliness. He will accept his status of being in the minority and will find courage to say 'yes' and 'no'. He will become a truly responsive individual.

These two assumptions also free the individual from the fear of authority. He will learn to experience a sense of authority in his own self. This, to us, is most important. Unless the individual in our society vests authority in himself, the systems in our society will continue to be one-person empires. It is only with the discovery of authority within the self that an individual can learn to treat himself as a representative in the system. Such an acceptance can generate a sense of responsibility toward the self and the system. Freedom from the fear of authority outside and acceptance of the authority from within can alone generate the responsibility for being a representative in the system. The agrarian ethos prevalent in India today denies the simultaneity of authority within and without. This simultaneity of location of authority is a dire need today. We suggest that the two assumptions of the self—I have no certain knowledge but I have no doubts, and I am that and more—are the most significant forces for changing the transaction system in our society. Without these definitions, the individual cannot learn to be responsive. He will continue to dissipate his energy in fretting and fuming and in occasional rebellion. He will continue to carry the burden of reactive feelings. Just as the individual has to discover that he is the best resource of society, he has also to discover that he is the best resource for himself and his roles in the system. He cannot actualise this discovery unless he accepts these two assumptions.

Understanding the Meaning of Responsive Intervention and Criticism in Action

In order to understand the meaning of criticism in action, let us first explore the phenomenology of transaction where two sets of feelings are aroused. One set is made up of reactive feelings which they are anchored in a unidirectional concept of the role. These feelings are at, toward, for, against and (perhaps) with the other individual. The other set of feelings are at the level of the self, which may be quite incongruent with the reactive feelings. For example, when an individual is angry at others he may feel helpless at the level of the self. Our experience with our society suggests that socialisation in the agrarian ethos has made the individuals largely blind to feelings at the level of self. They have been left with no option but to live in the world of reactive feelings and transactions. Disowning feelings at the level of self and remaining entrenched in reactive feelings turns individuals into objects. All their interactions also convert others into objects.

As the first step, criticism in action implies identification and understanding of feelings at the level of self, which lie buried under the massive load of instantaneously aroused reactive feelings in a transaction. As the second step, it requires the understanding of the trap of the reactive feelings. As the third step, it requires the understanding that the holders of dominant and significant roles of the system have trained themselves well in dealing with reactive feelings. They would be completely nonplussed if an individual were to state the feelings at the level of the self without bitterness and anger. As a fourth step, it implies identification and understanding of feelings at the level of self in the other individual. Naturally, this helps to separate the reactive feelings of the other individual. As a fifth step, it implies the responsibility of directing one's transactions toward feelings at the level of self of the other individual as well as one's own. A close examination of these steps will suggest that these responses can free the individual from objecthood and restore the dignity of being a subject. He will also provide the other individual the dignity of being a subject in his transactions.

Another aspect of the phenomenology between the self and the system involves expectations and counter-expectations. The expectations are varied in nature. Most often, we have expectations in terms of social obligations. However, at another level, the individual continues to nurse psychological expectations of support, acceptance, dignity, respect and a sense of wellbeing. The fragmentation of expectation into two sets, the social and the psychological, creates a great deal of confusion in the choice of action and response. If the individual responds with a sense of social obligation, the receiver of action does not feel accepted. We found from the participants that when their parents did something as a part of their social obligation, they received the product of their action but they did not experience the psychological gift of responsive action. It left them feeling cheated. Similarly, they felt that when they did something out of a sense of social obligation they did not feel a sense of participation or psychological involvement in the action.

Entrenchment in reactive feelings and acting from them transforms all people and situations into mere replicas of the primary system and the past. Reactive feelings are the primary source of transference of the family modalities to all secondary systems. Thus all interactions become pseudo-transactions. The phenomenology of transactions in the Indian situation ignores the present, the people and the issues. The whole ethos of living in transaction by reactive feelings is even immortalised by Indian agrarian society in the statement, '*Satyam broyat priyam broyat, na broyat apriyam satyam* (Truth can be spoken pleasantly, pleasant things can be spoken, but unpleasant truths may not be spoken).'

Criticism in action is nothing but a natural corollary of responsive intervention. An individual who intervenes in a responsive fashion presents a model of behaviour which is beyond the response hierarchies set by the prescriptive agrarian ethos. This act stands out in contrast to the present response hierarchy and acts as a criticism in action. It carries with it evidence of his freedom, maturity, responsibility and individuation. We believe that criticism in action is more effective than verbal or social criticism. As already shown, a study of messiahs and charismatic leaders like Gandhi, Christ and Buddha would reveal that their major instrument for social change was criticism in action, backed by responsive intervention at the feeling level. In today's times there is no alternative but to be our own messiahs in our own microcosm of social existence.

Would we dare? Or would we let the phantoms of loneliness, the accusing eyes of isolation, the pointing fingers of guilt and anxiety, and the voices shouting 'rebel' frighten us back into a morbid existence? We leave you here to ponder over the issues and make a choice. Let us offer you a summing-up in the following section.

Summing-up

Today in India and perhaps in many other Afro-Asian societies educated people seem to invest a large part of their time and energy in arguing about, discussing and propounding assessments of their societies. This assessment is in terms of political, economic and technological developments, as well as the nature and processes of democracy, social justice and equal opportunities, as indices of a society's political maturity. Formal forums and informal meetings are held to discuss the rate of growth, per capita income, capital formation, industrial expansion, standard of living, percentage of literacy, etc., in order to evaluate progress in terms of the standards of the West.

Simultaneously, these discussions and arguments tend to focus on corruption, black money, moral and legal deterioration, and lack of planning. Most discussions move towards identification of responsibility for failure. Depending upon the forum, the blame is put either on political leaders, the public or bureaucrats. The new generation is also condemned for its lack of commitment. The blame is finally laid on the cultural institutions of society. To many, the preoccupation with political ideology, socio-economic wellbeing and technological development is a sign of healthy concern for the state of the nation and society. While we admit that these concerns are legitimate and necessary, we are also convinced that an exclusive focus on them leads to empty slogans, self-hate and self-defeat.

It seems that the Afro-Asian societies, specially India, are seduced by the model of input–output progress. Exclusive obsession with this model is a festering disease which saps the energy of the nation. What are believed to be the basic foundations for a new society are nothing but a superstructure without foundations which collapses sooner or later. A new society cannot be ushered in through political, economic or technological manipulations alone. To attempt to do so is to chase a mirage. It can only create the dynamics of opportunities, exploitation and self-centredness. It keeps comparativism alive. Afro-Asian societies need more serious investment in the socio-psychological and psycho-cultural infrastructures of action, so that the people can carry these societies forward. Our study of the current generation suggests that Indian society, by investing time, energy and resources primarily and exclusively in political, economic and technological development, is preparing a fertile ground for the forces of disintegration and is facilitating the emergence of a despotic leadership.

Most Afro-Asian societies, whose developmental perspectives are anchored in the economic, political and technological models from the West, have thrown up and continue to throw up unmanageable dysfunctionalities in the sphere of social living and cultural identity. Most of them at some point have drifted into governance by autocratic individual or military juntas. Consequently, the nations have become barriers to their own growth. India, except for the period of Emergency, is one of the very few, perhaps the only, developing countries which has not been so trapped. The imposition of Emergency can be understood as an outcome of overemphasis on an eco-political ideology and models of technological growth from the West, combined with a deliberate neglect of investment in psycho-cultural infrastructures of action.

The exclusive commitment to Western models distorts the priorities of these societies and promotes the socio-cultural imperialism of the West. This process is more destructive than the direct eco-political imperialism of the West. In most of these societies the colonial rulers left the socio-cultural fabric of society alone. A serious research would suggest that they redeployed the socio-cultural fabric very effectively for their own ends. For example, the British in India used caste and parochial identities in the regimentation of the Indian army. Thus, they ensured that the army personnel could not, in real operational terms, become a unified body. The political masters restricted their role to legal and political control. For successful economic exploitation of native resources, they built efficient economic infrastructures such as transport and communications. They imposed an alien education system for creating appropriate manpower. It is not surprising that a sizable majority of the elite, having inherited the British legacy, continue to operate within the colonial model of development and ignore the building of psycho-cultural and psycho-philosophical infrastructures of integration and action. India in a political sense is a democracy, but in the administrative sense it is still a colonised country. We suspect this is true of most Afro-Asian nations.

Our study suggests that these societies will continue to plunge toward national disintegration unless investment in the reconstruction of their psycho-cultural institutions is planned. It is through these institutions that people can learn to internalise the ethos which will replenish the social system instead of only consuming it.

There is exclusive focus on the economic, political and technological development. Western experts attribute Indian problems to seven ills—religiosity, caste, superstition, orthodoxy, poverty, illiteracy and overpopulation. For the past sixty to seventy years the Indian elite has been merely echoing this analysis. Accordingly, they prescribe secularism, anti-casteism, rationalism, economic development, education and population control. Enlightenment for progress and development of a national identity remain incomplete goals. The credo of economic, political and technological development has led to very visible material results. India can really be proud of its economic and technological achievements. However, Indian society suffers from the pathology of caste, orthodoxy, religion and many other aspects. None of these programmes have led to investment in infrastructures for human beings. The Indian elite, with its Western cognitive maps, has not been able to generate any movement which can create a synergic process in Indian society. Contrary to the synergetic process, disruptive processes, accompanied by scepticism, hate for the system of belonging, violence, and disowning responsibility have become more intense. In the moment of economic success, social disintegration is quite evident.

The preoccupation with the ills of society has trapped the elite into the theory of change by the eradication of social ills through legislation. It is apparent that these action programmes are counterproductive, making people hold on to their 'ills'. In addition, hatred for one's own system of belonging is intensified. The two most potent weapons of destruction today, self-hatred and ambivalence, are ignored. Handicapped by Western blinkers, the intellectuals cannot design sufficiently strong new themes, actions, and the accompanying infrastructures to draw cultural and social energy away from the seven ills. Removal of these ills alone is meaningless, because their prolonged presence in the body of society has destroyed the social processes and institutions of regeneration and reformulation. Concerted effort and investment in the development of psycho-cultural institutions and infrastructures of social action are essential for a wholesome revival of society and for recapturing its dynamism.

The exclusive focus on economic, political and technological development has created a revolution of rising expectations. The suppressed pathos of deprivation, discrimination and denial, held in abeyance so far, has been awakened. People have responded to the economic, political and technological inputs with a singularly self-centred and consumer-oriented stance to life. Though they are aware of the need to replenish the resources of the system and to regulate their utilisation, they have conveniently pushed this responsibility on to the elite and its governance. The elite has then raised the cry of greater need for controls. They are sold to the idea that India needs an authoritative leadership, perhaps a benign despot. They have assigned the task of disciplining the people to themselves. And as people are manifestly not taking the responsibility, the elite have converted the need for control into a need for coercive force.

Socio-cultural and intellectual imperialism, the internalised ethos of the elite of India and perhaps of other Afro-Asian societies, has led to the deployment of national resources in sectors of economic development alone. This deployment is necessary, but the priorities chosen and the plan of deployment are not specific to the Indian situation.

It is really surprising that while the elite is perhaps as aware as any of us about the lack of socio-psychological and psycho-cultural mobilisation, it has made no effort to think of ways and means to promote a new action ethos. It has remained content with sermonising, advertising and cajoling. One wonders why the elite is averse to taking risks and to venturing into innovative socio-psychological and psycho-cultural processes and institutions. It is mortifying to see institutions and organisations systematically ignore all cues from the internal and external environment indicating that appropriate ways for a dynamic and mutual relationship between the individual and society need to be fostered. However, the commitment to achieve physical targets is so overwhelming that institutionalisation of processes necessary for that achievement are ignored. The only processes acceptable to the elite are control and coercion.

The present model continues to favour techno-informative education. The model has its advantages and can, as it has done, produce significant results without creating a sense of pride in the nation. The processes of development will remain geared to exploitation of opportunities created by the government. They will hardly be self-generative. The social space of living will become a seething mass of negative experiences. The social pathology of dowry, violence, parochialism and accompanying ills will keep pace with economic growth.

It is time that space is made available for the individual to resolve the following thought and feeling modalities:

No situation is my situation.
It is always somebody else's situation.

It is the situation of significant others.
They have to decide.
They have to make the decision for me.

Our study of the current generation quite clearly establishes that forceful efforts have to be made urgently to invest in individuals' socio-psychological and psycho-cultural infrastructures of action. The current generation just hankers for inputs. These will have to be supported and reinforced later by the larger society.

Even if it achieves nothing more, we hope that this book of ours will bring home to the nation the fact that the need of the hour is social, cultural and philosophical regeneration. Political, economic and technological development alone is not enough. The call for social regeneration is not new. Right from the days of Buddha to Gandhi, and now from the mouth of every political leader, this call has been echoed. Buddha and Gandhi, besides giving the call, took social action. The current leadership only seems to ensure that the social and institutional processes responsible for social degeneration continue to flourish.

However, our call is different from that of the saints. In their call there was nostalgia for India's past, especially with Gandhi, Dayanand and Raja Ram Mohan Roy. They wanted to fight degeneration by reanchoring Indian society in the psycho-philosophical ethos of the Upanishads and Puranas. Their attempt to integrate alien influences was selective and included only those elements which could be grafted on to the agrarian ethos. The regeneration of Indian society cannot be accomplished by reverting to the agrarian social design and ethos of the past. A new step has to be taken. To identify this step and design social action, an acute awareness of the psycho-cultural perspective, anchored in the internalised reality of the current generation, is necessary.

In this chapter some of the initial boundary conditions necessary to lay the foundations of socio-psychological and psycho-cultural infrastructures are described. These are not comprehensive and have been derived purely from the logic of the primary data provided by the participants. We hope that readers and other thoughtful people will engage themselves with these ideas in order to discover more cogent perspectives and action models.

We are sure that our reconstruction of Indian reality from the experiences of the current generation will not be validated by historians. Our interpretation of the ethos of India may not be fully substantiated by the interpreters of the scriptures. It may not agree with the hard-core political and economic facts and ideological stances taken by the elite. However, we are convinced that we have presented here, to a very great extent, the psychological and existential meaning of Indian reality experienced in the core identities of the current generation. It is the conviction that our reconstruction is rooted in live data that gives us the courage to bring this book to you. We will consider our task well-done and our efforts worthwhile, if it triggers off new reflections about the nature of Indian society. If it provokes readers to re-examine the various reconstructions available and then ask why they have not been able to generate dynamism, our reconstruction would have served its purpose. If it opens new doors and leads to responsive actions, the courage of thousands of participants to process themselves and to share their being in public will have earned its reward.

GLOSSARY

Abhivandan: Salutations, obeisance, praise.

Abrogation of representatives: A deliberate disowning and surrender of one's quality of membership in a group by becoming a follower and hence not being able to influence the course of group choice and action.

Acharyas: Religious instructors or spiritual guides, often sages.

Action space: The construct implies that in all transactions there is space for action. This action is not necessarily the normative, prescriptive one. Individuals have the freedom to assess the situation and design their own action.

Avatar: Incarnation, the descent of a deity or released soul to earth in bodily form.

Babas: Ascetics.

Bali: In the *Ramayana*, the great epic, Bali was the king of monkeys and brother of Sugriva. His son was Angade. He was slayed by Rama.

Bhasmasura: A rakshasa. He was given the boon of turning anything into ashes with the touch of his fingers. He wanted to test this on Shiva himself. Shiva ran around the world to escape from Bhasmasura. Ultimately, Lord Vishnu came in the form of a dancer and started dancing in front of Bhasmasura. Captivated, Bhasmasura also started dancing with this dancer. In one of the postures, the dancer put her hands on the head. Bhasmasura also imitated this action and was immediately turned into ashes.

Catharsis/Catharses: Evocation of withheld emotions and feelings and creating an experience of relieving the stress associated with them. Sometimes it is simply an opportunity for purging oneself of such emotions in public. Primarily catharsis is a process which encourages and permits individuals to discharge pent-up, socially unacceptable affects.

Cognitive map: Frames of perception largely anchored in the construction of reality around oneself. The logic of the construction of this reality is learnt through growth and is very often determined by systematic exposure to values, definitions of actions, meanings of situations and meanings of people and objects in one's community and society. For example, for a tribal man the first exposure to a policeman standing on a traffic island and controlling traffic has no meaning. Such a situation and hence experience does not exist.

Coherence: Coherence is a postulate and a process. The postulate of coherence is that any system has diverse elements which are discrete. They are also discontinuous and simultaneous. Some of them may be analogues to each other and others may be contradictory and opposite. All these elements put together make a balanced system. Heterogeneity with a rhythm and an order is the nature of any system. The process is to sustain this diversity and heterogeneity. There is a general tendency in human systems to create an artificial homogeneity and analogous element. The entry of heterogeneity is denied, masked or registered. Very often human systems treat this diversity as a problem and waste their energies in managing it. The ancient Indian design of social structure is a classical example of creating coherence. The construct of coherence is best illustrated in the metaphor—diversity in unity and unity in diversity.

Congruence: This postulates that processes of identity, society and culture need congruence to create effective settings for human growth. For example, meaning-making as a common process of identity, society and culture. The very fact that each human action, choice or input brings about changes in social and cultural systems, creating a feedback loop to identity, creates variance in the meaning-making processes of the three systems. Congruence is then a process by which individual or groups try to recreate a consonance between the meaning-making processes of the three systems.

Congruence of microcosm and macrocosm: Microcosm and macrocosm have many similarities. However, in each moment the individual is a microcosm and he operates in a macrocosm of the Thou. The macrocosm and the microcosm in their independent unfolding may come together where their locations are not congruent. This inevitably creates stress, stalemates and, often, major confrontations. Achieving congruence is one of the most significant human endeavours.

Convergence: It postulates that the elements of the three systems of identity, society and culture act as vectors in a field force. The forces behind the elements of these systems tend to act in different directions and often tend to contradict, deny or neutralise each other. This creates problems of decision-making, choice-making and role taking for individuals. Society goes through processes of anomie. Convergence is then a process by which the forces of each elements are made to converge so that they create a synergy for transactions of replenishment and wellbeing between the systems of identity, society and culture.

Correspondence: Correspondence is a construct derived from three sources. One is from the correspondence theory in philosophy; the other is the mathematics function of correspondence; and the third is the usual connotation of similarity and agreement existing between two elements. Here, the construct conveys that corresponding elements exist in the human identity, the social system and the cultural system. It also connotes a process of bringing these elements in the three systems to awareness and making them a basis for making choices for relevant action, so that the balance between identity, social and cultural systems is optimised.

Emotive/Cognitive life space: The construct of life space is that each individual in this process of being and becoming acquires a location. This location is grounded in a world-view, a mythology and a saga of being a member of a class. Consequently, a space for mobility, aspiration and fulfilment gets defined. Its boundaries are often vague. However, on the whole, the space is fairly stable. Within this space there is a cognitive sector and an emotive sector. The latter is rooted in the sagas and myths. It is also coloured by the emotional experiences of living and growing in the current times. The cognitive sector is well-defined by normative codings devised by the elders. It is also reinforced by interactions with members of other classes. The emotive space is the space where dreams, wishes and aspirations along with frustration, doubts and anxieties operate. It is the space of pathos. The cognitive space is the space for logical action and acts which are in keeping with survival issues in society.

Erotic feeling: The word here is used to encompass feelings relating to attachment, attraction and a wish for togetherness—psychic and psychical. It is not used in its restricted meaning of sexuality. The word connotes here the Sanskrit word '*Kama*' and not the word '*Rati*'.

Event structure: This construct was formulated in 1965 to describe the mode of analysis generally utilised by social scientists who follow the framework of physical sciences to understand human events. In this analytical mode they focus only on identifying the cause–effect relationship of 'what led to what' and 'why'. It is a kind of manifest analysis of the event without taking the contextual data into account.

Existential: A term from the social sciences, it has wide connotations. It is a derivative of the philosophy of existentialism which states that man forms his essence in the course of the life he chooses to lead. The doctrine implies responsibility on man for shaping his own nature. It emphasises the importance of personal freedom, commitment and personal decision.

Existential themes: Transactions in a human system can be viewed from a normative perspective or an existential perspective. Sets of transactions can be analysed to discern a pattern, in which there could be a transactional theme, a normative theme and/or an existential theme. The existential theme then relates to the underlying struggle of a set of individuals to establish meanings for their own beings in their transactions with others.

Gaudhuli: A Sanskrit word for dusk. In the agrarian Indian society it got associated with the return of the cows who raised dust and masked the spread of colours left by the setting sun.

Grihasta/Karta: Householder, head of the family.

Gurudevas: Great teachers or spiritual guides, often sages.

Gurukul: A residential teaching school run by a tutor.

Individuation: Individuation is a social science construct. It has been in use from the times of early psychoanalysts. It entered mainstream social sciences in the late 1960s. Essentially it means the emergence of an individual from the general mass of human beings. In its special sense it is an active process of an individual by which he defines his location with regard to social issues. Such an individual neither feels the compulsion to conform nor gives into alienation from the society. He persists and experiences his membership right to define and make choices in a social situation where norms have become matter of controversy and confusion.

Introjection: A psychoanalytical concept introduced by Frenczi. According to him, during the first one and a half years the child is centred around his oral processes. He puts things into his mouth and ingests and incorporates them within. This is not with regard to food alone but every other input such as feelings and experiences from other persons. These 'introjected' elements become an integral part of the foundation of his developing character and identity.

Macrocosmic quality: This term has been used to describe that nature of experience which acquires the immense influence of the whole universe, in contrast to the small living space of man.

Macrodynamic: Borrowed from physics which deals with the motion of large and overdeveloped bodies in the universe.

Maharishis: Great sages or saints.

Maya (Transience): A phenomenological word of sanskrit origin. Maya varies in its meanings, from a stream of consciousness, ever changing reality to what in decadence period people call the illusion of life. Essentially, maya is the eternal rhythm of life and the issues human beings confront is, 'What stance to take?' In being a part of this maya most societies for their own sense of security and perpetuation deny the flow of life and train people to live by norms which may not be relevant for the current flow of life of society.

Microcosm: Microcosm is used in the same sense as it is used in philosophy and physics. An atom is a microcosm of the whole universe of matter. Its processes are assumed to be identical with the processes of the whole universe.

Monolith: Monolith originally means a single block of stone, usually very large, around which the architecture and sculpture is designed. Here it is used in its derivative sense, where a person is turned into a larger-than-life image and the whole ethos and system are centred on his values, beliefs, etc. Many Indian charismatic leaders have acted as monoliths, for example Nehru.

Moratorium: A construct used for the first time in psychoanalytic literature to cover many psychological states of adolescence. This is the period during which an individual re-evaluates his childhood and assesses the forthcoming reality of adult life to give himself an identity. There is a rich psychological life which involves experimentation, exploration, questioning and reflecting before the individual makes his choices. One of the things we believe is that today's young generation is being pushed into achievement and action so relentlessly that there is not enough time for them to go through this essential process of moratorium.

Ontogenetic: A term derived from the field of biology and evolution, ontegeny stands for the course of development of an individual. Its parallel is the word 'phylogeny' which stands for the evolution of a race, a genus of plants, etc.

Organismic: This is word derived from the philosophy of organicism. According to the theory all vital activities of an organism arise from its autonomous composition and not from any one dominant part. Accordingly the character of the whole becomes the final determinant of behaviour of the parts and its composition. In one way organismic is a part of the vocabulary of the 'holist', 'gestalt' and 'existentialism'.

Phenomenological: Phenomenological is a derivative of the word phenomenology, which is a philosophy concerned with phenomena. It stresses a careful description of phenomena in all areas of experience. Here, phenomenology is extended to imply being in contact with the living rhythm and unfolding of life in each moment of time and space.

Pravara: Eminent, chief, exalted.

Primary props: A construct derived from the world of theatre. Primary props provide the stage and the setting in which the drama unfolds. Here, the word has been used in the same sense, in that every presentation of the self and identity has props around which it unfolds.

Puranas: Sacred writings in Sanskrit on Hindu mythology, folklore, etc.

Samskara: Purification, refinement, ornamentation, conservation.

Subversion of the system: Subversion of a system is always an internal process by which its own members tend to create distortions in the processes of the system concerned with goal achievements. It can be active or passive. Active subversion is often expressed by actions such as stopping work—tools down, slow down. Passive subversion is more subtle. It is expressed in silent collusion, through which the sources of organisation are wasted. This may involve delays in decision-making, routing files in a wrong way, not attending to visible indiscipline and many other such phenomena.

Tertiary source: Three times removed from the original source. Another meaning is the third order manifestation of a primary phenomenon.

Therapeutic community: A concept promoted by psychiatry. The belief here is that mental health issues are not centred on the psyche of the individual alone. Mental sickness according to this view is a burden that some individuals carry as part of the stress and ill health of processes in the community, i.e. a family, a neighbourhood, a larger segment of the population. It maintains that instead of mental asylums, communities need to be created which are organismic and provide a context of mental wellbeing for people.

Transactional themes: One way of conceptualising the human system is to talk of the set of transactions that occurs between its components at various points over a period of time. A phenomenological description of these transactions often leads to an identification of patterns of transactions, which revolve around certain themes of relatedness. These are called transactional themes.

Transcendental: The word is derived from transcend. According to the dictionary it implies rising above the bounded normative meanings, perspectives and views.

Transience: This construct was formulated by the authors in 1984. It defines that process and state where conflicting ethos of diverse origins prevail in society simultaneously. The condition creates a struggle in the individual as well as social systems about choices of behaviour. Definitions and guidelines of ethical choices get disturbed and a process for the emergence of a new ethos begins.

Unit based social organisation: The term is used to describe the Indian social organisation where governance of communities was left to the communities themselves. There was no centralised governance carried out from a distance. In this unit based social organisation the lifestyle of each village or a set of villages evolved from its own basic socio-cultural and geo-physical resources.

Upanishads: Philosophical compositions concluding the expositions of the Vedas.

BIBLIOGRAPHY

Alvares, Claude Alphonso (1979), *Homo Faber: Technology and Culture in India, China and the West*, Bombay, Allied Publishers.

Anand, Mulk Raj (1963), *Is There a Contemporary Indian Civilisation*?, Bombay, Asia Publishing House.

Atreya, Bhikhan Lal et. al (1963), *Indian Culture*, Delhi, Universal Book and Stationary Co.

Basham, A.L. (1963), *The Wonder That was India: A Survey of the Culture of the Indian Sub-continent Before the Coming of Muslims*, Calcutta: Orient Longman.

Birnbaum, Norman (1969), *Man in the Age of Technology*, New York, Columbia University Press.

Bishop, Donald H., ed. (1982), *Thinkers of the Indian Renaissance*, New Delhi, Wiley Eastern Ltd.

Breman, Jan (1979), *Patronage and Exploitation: Changing Agrarian Relations in South Gujarat, India*, New Delhi, Manohar.

Brunton, Paul (1976), *A Search in Secret India*, Bombay, B.I. Publications.

Bose, N.K. (1967), *Culture and Society in India*, Bombay, Asia Publishing House.

Broomfield, John H. (1968), *Elite Conflict in a Plural Society*, Bombay, Oxford University Press.

Buck, Pearl S. (1982), *Good Earth*, New Delhi, Oxford University Press.

Chakraborty, Chandra (1968), *Cultural History of the Hindus*, New Delhi, Deep and Deep.

Chakravarti, K.C. (1961), *Ancient Indian Culture and Civilisation*, Bombay, Vora & Company Publishers.

Chattopadhyaya, Debiprasad (1967), *Individuals and Societies: A Methodological Inquiry*, Bombay, Allied Publishers.

Chattopadhyaya, Sudhakar (1961), *Traditional Values in India Life*, New Delhi, India International Centre.

——— 1965, *Social Life in Ancient India*, Calcutta, Academic Publishers.

Chaudhuri, Nirad C. (1964), *The Autobiography of an Unknown Indian*, Bombay, Jaico Publishing House.

Chopra, Premnath (1963), *Some Aspects of Social Life during the Mughal Age*, Agra, Shivlal Agarwal & Co.

Datta, Kalikinkar (1964), *Dawn of Renascent India*, Bombay, Allied Publishers.

———(1965), *Renaissance, Nationalism and Social Changes in Modern India*, Calcutta, Book land.

Dixit, Prabha (1974), *Communialism: A Struggle for Power*, Bombay, Orient Longman.

Dwivedi, O.P. (1989), *World Religions and the Environment*, New Delhi, Gitanjali Publishing House.

Garg, Pulin K., and Indira J. Parikh (1976), *Profiles in Identity*, New Delhi, Vision Books.

———(1980), 'Trishanku Complex: A study of MBAs Passage to Adulthood', Research Monograph, Ahmedabad, Indian Institute of Management.

———(1988), 'Values, Design and Development of Strategic Organizations', in Pradip N. Khandwalla, ed., *Social Development: A New Role for the Organizational Sciences*, New Delhi, Sage.

———eds. (1990), *Organization Design and Definitions Proceedings of the International Conference on Transience and Transitions in Organizations, Vol. III*, ISISD Publications.

Gehlen, Arnold (1980), *Man in the Age of Technology*, New York, Columbia University Press.

Ghai, O.P. (1986), *Unity in Diversity, A Guide to the Understanding of the Fundamental Unity Underlying the Great Living Religions of the World*, New Delhi, Sterling Publications.

Ghose, Sisirkumar (1977), *Man and Society: As on a Darkling Plain*, New Delhi, International Book Corner.

Ghurye, G.S. (1963), *Anatomy of Rururban Community*, Bombay, Popular Prakashan.

———(1986), *Caste and Race in India*, 5th ed., Bombay, Popular Prakashan.

Gopal, Subramania, (1975), *Outlines of Jainism*, New Delhi, Wiley Eastern Ltd.

Gupta, Giri Raj , ed. (1983), *Religion in Modern India*, New Delhi, Vikas.

Heesterman, J.C. (1985), *The Inner Conflict of Traditions: Essays in Indian Rituals, Kinship and Society*, Delhi, Oxford University Press.

Jain, S.P. (1975), *The Social Structure of Hindu-Muslim Community*, New Delhi, National.

Karandikar, Maheshwar Anant (1968), *Islam in India's Transition to Modernity*, Bombay, Orient Longman.

Karunakaran, K.P. (1965), *Religion and Political Awakening in India*, Meerut, Meenakshi Prakashan.

Karve, I. (1961), *Hindu Society: An Interpretation*, Poona, Deccan College.

———(1968), *Kinship Organisation in India*, 3rd ed., Bombay, Asia Publishing House.

Lacy, Geighton (1965), *The Conscience of India: Moral Traditiona in the Modern World*, New York, Holt, Rinehart and Winston.

Lannoy, R. (1971), *The Speaking Tree: A Study of Indian Culture and Society*, London, Oxford University Press.

Larsen, Gerald James (1990), '*India Through Hindu Categories*.: A Samkhya Response', *Contributions to Indian Sociology, 24:2*

Macy, Joanna (1985), *Dharma and Development, Religion as Resource in the Sarvodaya Self-help Movement*, rev. ed., West Hartford, Kumarian Press.

Mahajan, M. (1986), *Development of Material Culture in India*, Delhi, Sundeep Prakashan.

Majumdar, R. C., ed. (1951), *The History and Culture of the Indian People*, Bombay, Bhartiya Vidya Bhavan.

Malik, Yogendra K. (1990), *Boeings and Bullock-Carts: Studies in Change and Continuity in Indian Civilisation*, Delhi, Chanakya Publications.

Mishan, E.J. (1969), *Growth: The Price We Pay*, London, Staples Press.

Mitra, Sisirkumar (1963), *Resurgent India*.

Moddie, A.D. (1968), *The Brahmanical Culture and Modernity*, Bombay, Asia Publishing House.

Moraes, George Mask (1964), *A History of Christianity in India, from Early Times to St. Francis Xavier, AD 52–1542*, Bombay, Manaktalas.

Mowat, R.C. (1987), *Climax of History*, New Delhi, Anmol Publications.

Munshi, K.M. (1962), *Foundations of Indian Culture*, Bombay, Bhartiya Vidya Bhavan.

Mukherjee, Radhakamal (1951), *The Indian Working Class*, 3rd ed., Bombay, Hind Kitab.

———(1964), *The Dimensions of Values, a Unified Theory*, London, Allen & Unwin.

———(1964 a), *The Sickness of Civilisation*, Bombay, Allied Publications.

———(1964 b), *1889—The Destiny of Civilisation*, Bombay, Asia Publishing House.

———(1966 c), *The Community of Communities*, Bombay, Manaktalas.

Murphy, Gardner (1953), *In the Minds of Men: The Study of Human Behaviour and Social Tensions in India*, New York, Basic Books.

Nehru, Jawaharlal (1964), *Discovery of India*, Bombay, New York, Asia Publishing House.

Parikh, Indira J., Vipin K. Garg, and Pulin K. Garg, eds. (1988), Proceedings of International Conference, *Transience and Transitions in Organisations: Volume I Corporate Culture of India (Perspective of Chief Executives)*, ISISD Publications.

Parikh, Indira J., and Pulin K. Garg (1990), 'Indian Organisations: Value Dilemmas in Managerial Roles', in R.N. Kannungo and A.M. Jaeger eds., *The Management of Organisations in Developing Countries*, London, Rutledge and Kegan Paul.

———(1989), *Indian Women: An Inner Dialogue*, New Delhi, Sage.

Radhakrishnan, S. (1958) *Eastern Religions and Western Thought*, London, Unwin Books.

———(1961), *An Idealist View of Life*, 2nd ed., London, Unwin Books.

———(1965), *The Hindu View of Life*, London, Unwin Books.

Radhakrishnan, S., and Charles A. Moore (1975), ed., *A Source Book in Indian Philosophy*, Princeton, Princeton University Press.

Radhakrishnan, S., and Raju P.T. (1966), *The Concept of Man: A Study in Comparative Philosophy*, 2nd ed., London, Allen & Unwin.

Ram Gopal (1963), *British Rule in India: An Assessment*, Bombay, Asia Publishing House.

Ross, Nancy Wilson (1966), *Hinduism, Buddhism, Zen: An Introduction to Their Meanings and Their Arts*, London, Faber and Faber.

Sen, Kshitimohan (1961), *Hinduism*, Harmandsworth, Penguin.

Sen, S.P. (1979), *Social and Religious Reform Movements in the 19th and 20th Centuries*, Calcutta, Institute of Historical Studies.

Sharma, Shripad Rama (1964), *Our Human Heritage: A Synoptic Study*, Bombay, Bharatiya Vidya Bhavan.

Sircar, D.C. (1967), *Studies in the Society and Administration of Ancient and Medieval India*, Calcutta, Firma KLM.

Snyder, Louis Leo (1967), *The Making of Modern Man: From the Renaissance to the Present*, Princeton, Van Nostrand.

Sorokin, Pitirim Aleksandrovich (1962), *Social and Cultural Dynamics*, New York, Bedminster Press.

Stanley, Manfred (1972), *Social Development: A Critical Perspective*, New York, Basic Books.

Stavenhagen, Rodolfo (1975), *Social Classes in Agrarian Societies*, trans., Judy Adler Helmann, Garden City, Anchor Press.

Stearns, Peter N. ed. (1972), *The Impact of Industrial Revolution: Protest and Alienation*, Englewood Cliffs, Prentice Hall.

Swamy, Satprakashananda (1965), *Methods of Knowledge, Perceptual, Nonperceptual and Transcendental According to Advaita Vedanta.*

Thapar, Romila (1978), *Ancient Indian Social History: Some Interpretations*, New Delhi, Orient Longman.

Theodore, Rozak (1969), *The Making of a Counter Culture: Reflections on the Technocratic Society and its Youthful Opposition*, Garden City, Doubleday.

Unnithan, T.K.N., ed. *Towards a Sociology of Culture in India: Essays in Honour of Prof. D.P. Mukherji*, New Delhi, Prentice-Hall of India.

Vyas, K.C. (1957), *The Social Renaissance in India*, Bombay, Vora Publishers.

Webster, John C.B. (1976), *The Christian Community and Change in Nineteenth Century North India*, Delhi, Macmillan.

Weiner, Myron (1966), *Modernisation: the Dynamics of Growth*, New York, Basic Books.